Solar drying:
Practical methods of
food preservation

Solar drying: Practical methods of food preservation

Prepared with the financial support of the United Nations Financing System on Science and Technology for Development (UNFSSTD)

International Labour Office   Geneva

ISBN 92-2-105357-1

First published 1986

Printed by the International Labour Office, Geneva, Switzerland

PREFACE

The lack of technical and socio-economic information on alternative or improved food processing technologies available to people living in rural areas of developing countries means that much food gets wasted. Excess production in the villages is rarely preserved for times when fresh food is not available. Emphasis is more often placed on large-scale food processing complexes which frequently require only a few skilled operators, use imported equipment and packaging, and produce foods which are expensive and have a low nutritional content.

Sun drying of foods is a technique that has been in use for centuries, with little change in the methods employed. This frequently results in poorly dried, infested products. The use of improved sun drying techniques and the introduction of solar drying, by which means foods can be dried even in humid, cloudy climates, can greatly improve both the quality and quantity of goods produced and be of great benefit to people living in rural areas.

This manual, which is an outcome of an ILO-executed project in four-least developed arab countries (Sudan, Somalia, Democratic Yemen and the Yemen Arab Republic) to promote the choice, development and application of appropriate food processing technologies, aims to explain in easily understood terms how food drying techniques can be introduced or improved. In this way, surplus food can be preserved which can be used later to make nutritious meals. All the equipment or materials necessary can be obtained locally. The project was financed by the United Nations Financing System for Science and Technology for Development (UNFSSTD).

During the project it became apparent that there was a need to give extension workers basic technical information which could be readily assimilated and which could be directly applicable. In the case of solar drying most of the work is, at present, carried out at an academic level in research institutes, and little attention is paid to the application of the information in rural areas.

It is hoped that this manual will supply agronomists, engineers and food technologists working in this field with the basic theory and practice of sun and solar drying. With an understanding of national needs and priorities they should then be able to advise extension workers of appropriate applications of these techniques. Some general guidance is also provided on the type of information extension workers might need and the methods they could use to obtain this. It remains up to the English-speaking reader to translate these into a locally comprehensible form for local use.

The manual includes a step-by-step guide to building different types of dryers using locally-obtainable raw materials. Fish drying, vegetable and fruit drying and grain drying are all covered, and appropriate processing methods discussed. The necessity of good packaging is stressed. It is very important to ensure an adequate "shelf-life" of the product and to prevent its untimely deterioration.

Some references are given as sources of more detailed, supplementary information. Access to technical literature can be difficult but the reader is reminded that solar drying is of world-wide interest and there may be a national or regional institute working in this field, from which information can be obtained.

Each chapter can be interpreted at two levels. Most of the technical information supplied will be of interest to the technologist. A summary of the pertinent points which should be passed on, in a suitable form, to the extension worker is given at the end of each chapter.

Chapter 1 provides an introduction to the type of information which is required and the approach which should be adopted to establish the feasibility of a solar drying exercise.

Chapter 2 describes the basic drying theory and explains how the sun's energy can be harnessed to dry foods. Some basic solar dryers are described with some guidance on their methods of construction.

Chapter 3 discusses how the technologist can work with the extension worker to encourage the adoption of improved technologies by the rural people in developing countries.

Chapter 4 provides information on simple methods suitable for drying fish. Vegetable drying is discussed in Chapter 5. The preservation of fruit by drying is covered in Chapter 6 and grain drying is discussed in Chapter 7.

This manual was prepared by Dr. C.I. Speirs of the Tropical Development and Research Institute, London, in collaboration with Ms. H.C. Coote, staff member of the Technology and Employment Branch of the ILO.

A.S. Bhalla,
Chief,
Technology and Employment Branch

CONTENTS

ACKNOWLEDGEMENTS

The publication of this training manual on solar drying
was made possible by a grant from the United Nations
Financing System for Science and Technology for Development.
The International Labour Office acknowledges this generous support.

CHAPTER I

ELEMENTS IN THE CHOICE OF SOLAR DRYING OF FOOD PRODUCTS

I. Evaluation of solar drying potential

The preservation of foodstuffs by drying is believed to be one of the first food processing techniques used by man, developing in conjunction with the cultivation of food grains, in the Middle East. The traditional method of crop drying practised over the centuries throughout the world is sun drying, where the foodstuff is spread on a flat surface in the open air and exposed to the drying action of the sun. Variations on this technique include hanging the foodstuff from the eaves of buildings or from trees or gathering the harvest in bundles in the fields. Today, sun drying still remains the most widespread method of food preservation.

The success of the technique can be attributed to its simplicity and low cost. Under favourable climatic conditions good quality products can be obtained. However in an unreliable climate, losses due to spoilage can be excessive. In wet or humid weather moisture loss from the food can be intermittent and irregular and the rate of drying slows down. This increases the risk of spoilage and reduces the quality of the product. It is likely that some of the foodstuff will be overdried, while a portion may be unacceptably moist, depending on its location within the batch. Contamination by dust and infestation by insects is unavoidable. Birds and animals will consume some of the crop and also constitute another source of contamination. This creates an extra task: to remain vigilant in order to cover the crop in the event of rain or dust storms, and also to scare away potential predators in an attempt to control sun drying losses.

In the industrialised countries, the food processing sector is typified by high labour costs and increasingly stringent quality standards. One response to the problems associated with sun drying has been the development and use of high capacity, artificial drying plants capable of giving a high quality product irrespective of weather conditions. These plants are usually energy-, capital- and technology-intensive, and have a low labour requirement using mainly skilled process and maintenance personnel. The dehydration units are relatively inflexible and are typically geared towards a large throughput of a single product. Such processes are not generally suitable for the needs of the small-scale farmer in developing countries who produces small quantities of foodstuffs to be dried for short periods throughout the year.

Artificial or fueled mechanical dryers are used in humid tropical regions, largely in the equatorial rain forest belt where daily downpours are predictable and the skies are usually overcast. In these conditions, the potential for sun drying is limited. Such dryers are typically associated with the so-called plantation crops such as cocoa, coffee and copra, where the cost of the drying operation can be justified by the foreign exchange generated by the product. A source of energy is required and the usual fuels available are wood or charcoal. This requirement restricts the use of such dryers to forested areas where such fuel is abundant and the ecological damage caused by cutting the wood is minimal. In some cases it may be possible to supplement or replace the fuel with by-products from the process such as bagasse in the case of sugar or coconut shells in the case of copra.

In arid or semi-arid regions where wood stocks are low and may already be insufficient to meet the cooking needs of the rural sector, the most suitable solution to processing problems may be to improve existing sun drying methods or to introduce solar drying techniques.

Solar drying, where the principal source of energy is derived from the enhancement of the sun's radiation, can be an improved alternative to sun drying. Compared with sun drying, solar drying provides higher air temperatures and consequential lower relative humidities which are conducive to improved drying rates and a lower final moisture content of the dried crop. As a result, the risk of spoilage during the actual drying process and in subsequent storage is reduced. The higher temperatures attained inhibit insect and microbial growth. Drying in an enclosed structure has the additional benefit of providing protection against rain, dust, insects, animals, and birds. All these factors contribute to improved and more consistent product quality.

On first impression, solar drying may appear to be the ideal solution to many food drying problems. The devices are of simple design and can be constructed using a high local material content. The energy source is freely available and poses no waste disposal problems. However, it should be emphasised that the process is not always technically feasible, economically attractive, or socially desirable. Strenuous efforts must be made, in co-operation with extension agencies and other interested organisations, to determine as accurately as possible the nature and quantity of commodities that could be dried. The reasons for drying the selected foods and the required quality for the market outlets should be clear. An extension worker may be enthusiastic about introducing, say, solar pepper drying in his or her region, but without clear-cut reasons for doing so the project may be doomed to failure.

The extension worker should be aware that some building and maintenance costs will be involved, and it is obviously advantageous that the solar dryer be used for as long a period during the year as practically possible. It may be more cost-effective and socially desirable in some locations to use the dryer on a communal basis.

In the planning stage, therefore, the technologist should establish some of the facts listed below. Not all of the questions will be relevant to any one project, and time and manpower constraints may render a detailed investigation impractical. However, any information will be useful.

II.1 Estimation of commodity production

It is appreciated that it may be difficult to obtain detailed information on the quantity of the commodity that is harvested, particularly where the food is consumed by the producer or the producer's family and where there is no formal or centralised form of purchase or market. However, where possible, information should be collected on:

(a) the quantity of fresh material produced in the growing season by:
- each farmer/fisherman
- the organisation (e.g. cooperative) in which the farmer/ fisherman participates
- each district
- the country;

(b) the duration of the harvest season:
- for a farm/fishery
- within a district;

(c) the amount of the commodity harvested in a day;

(d) the likely increase or decrease in the production of the commodity in the near future.

II.2 Present drying practices

It should be established whether the food stuffs are currently dried, and if so, by what means. Traditional techniques such as sun drying or even artificial drying may already be used.

If alternative drying practices are being carried out then the following facts should be established:

- the amount of an individual farmer's crop which is dried;
- the nature of any processing carried out after harvest and prior to drying;
- the moisture content of the commodity before and after drying; or alternatively, the wet to dry ratio, i.e. the weight of the commodity prepared for drying compared to its weight when dried;
- the size, shape and other important features of the commodity prior to drying;
- the actual techniques used to dry the commodity. Every effort should be made to obtain this information as precisely as possible. If possible, the cost of this operation should be established;
- the problems experienced with these techniques, e.g. high capital or operating costs, high labour requirement, poor product quality etc.;
- post-drying processing operations carried out prior to sale or storage;
- the means of storage of the dried commodity before further processing, sale or consumption.

II.3 Product quality considerations

The quality of the dried product is of considerable importance. For dried fruit and vegetables sold to the local consumer, the main quality factor is the general appearance of the dried material whereas for commodities such as spices or pyrethrum, the content of extractable constituent is the main aspect of quality. For dried grain the moisture content is of particular importance. The importance of quality can be gauged from the following:

- the features of the dried product that determine its selling price, e.g. appearance, colour, size, shape, moisture content, purity, extractable constituent, degree of contamination, microbiological quality;
- the methods by which the quality factors are evaluated, e.g. by visual examination or laboratory analysis;
- variation of standards of quality for different markets;
- the relationship between product quality and selling price.

II.4 Markets for the dried commodity

As with any development of a new or improved product, knowledge must be gained at an early stage of the present market for the traditional product or the potential market for an improved product. Though such information may well be difficult to obtain in certain areas, particularly from rural communities, it is important that an attempt be made in order to determine the level of technology and the economic boundaries for the subsequent technical development of a solar dryer.

Information must be sought concerning the following:

- the (envisaged) outlets or markets for the dried commodity:

    (i)   self-consumption;
    (ii)   local sale;
    (iii) sale to large towns at some distance from the producer, either by the producer or via a third party;
    (iv)   export;
    (v)   further processing.

- consumer acceptability of the product. This is of particular importance when no dried product is currently available or is known to the potential market;

- marketing mechanisms or organisations for bringing producer and buyer/consumer together;

- the price currently obtainable for the fresh (or unprocessed) commodity. This may vary widely from season to season or from district to district;

- the price that can be obtained for the dried product.

There may be a possible alternative use for the fresh commodity. Surpluses of the commodity may be sold to a local entrepreneur for sale elsewhere, or another preservation operation carried out, e.g. pickling of vegetables, jam-making from fruits and so on. If this is the case, as much information as possible should be obtained on drying processes particularly with regard to their economic viability.

II.5 Project viability

By collecting the above information the technologist will be able to reach some preliminary conclusions about the potential viability of introducing solar drying. It may also be possible to draw conclusions about the attractiveness of solar drying relative to other processing operations.

The extension worker may be able to assist in the collection of more basic information, i.e. whether there is a glut of fresh produce or insufficient drying capacity at a particular location, and so on. This identification of a local need may in most cases justify developing a project.

At this stage it should be determined whether any applied research has been carried out, either nationally or in other countries, on the method for drying the commodity. Solar drying methods which have been tried and tested elsewhere may be directly applicable. It may be possible to modify techniques described for other commodities, or to adapt recommended artificial techniques to solar means. An existing sun-drying method could also be improved. A continuous interface between the technical organisation and the extension services should ensure against any unnecessary duplication of effort.

CHAPTER 2

DRYING THEORY AND PRACTICE

I. The mechanisms of drying

With the possible exception of grains, a complete understanding of the mechanism of drying of foods has not yet been developed. What information is available has often been obtained under isolated laboratory conditions using a single layer of the commodity as a test sample. Such sample drying behaviour will differ from bulk drying because the drying pattern in one portion of the batch will be affected by that in another portion of the batch. At its simplest this might mean that moisture removed from one grain will be absorbed by a second grain from which it will then be released to be absorbed by a third grain and so on before eventually being expelled into the air.

However it is generally accepted that there are two basic phenomena involved in the drying process: the evaporation of moisture from the surface, and the migration of moisture from the interior of a particle to the surface.

I.1 Surface evaporation

Moisture evaporates from a free surface on food in the same way as it evaporates from any free water surface, such as a lake. So long as the surface remains completely wet the rate of evaporation is constant. The principal factors affecting the rate of evaporation from a free water surface are the degree of movement of air over the surface and the temperature and humidity of the air.

Humidity is a measure of the moisture content of the air. Very dry air (i.e. low humidity) will have a greater capacity to evaporate water from a free surface than moist humid air. This concept of air having different capacities to evaporate moisture can be difficult to put across to a layman. One method of illustrating this point is to draw attention to the relative personal comfort of an arid inland area compared with a coastal strip at the same temperature. This is largely due to the dry inland air removing moisture produced by perspiration freely from the skin.

An increase in the air temperature will reduce the humidity, thereby increasing the driving force of the air to evaporate free moisture. While this brief description should suffice to outline the fundamental principle behind air drying the interested reader is referred to any standard text on psychrometry for a more detailed and exacting treatise.

The degree of movement of air over the particle surface is also of great importance. As moisture leaves the surface it passes to the air immediately adjacent to it. This increases the humidity of the surrounding air, thus reducing its capacity to evaporate more moisture. Therefore, unless the air surrounding the particle is replaced by fresh, comparatively dry air, an equilibrium will be reached between the particle and the air and no further evaporation will occur. In practice, this is rarely the case since even with very low natural convection currents there is sufficient movement of air. Increasing the air velocity (e.g. by the use of a fan) will markedly increase the rate of evaporation as the surface of the commodity will be in contact with relatively dry air at all times.

Most of the heat necessary for the evaporation of moisture from a food surface is supplied from the air by convection. However the transfer of heat to the food by conduction and radiation can also be important. Foods placed in solar dryers may be exposed to the radiation of the sun and if the food is placed on metal trays they will receive heat via conduction through the tray bottom.

I.2 Moisture migration

It is generally accepted that there are two principal mechanisms governing the migration of moisture from the internal structure of the food to the surface. These are diffusion and capillary flow. The most important factors affecting the rate of moisture migration are the temperature of the food particle, its moisture content, and the size of the piece. The higher the temperature of the food, the greater will be the rate of moisture migration. As the moisture content of the foodstuff decreases, the rate of migration will also decrease (since there is less moisture to migrate). The rate of migration will increase with decrease in particle size.

I.3 Drying rates

The drying rates of a foodstuff alter during the drying cycle reflecting the changes which are occurring in the composition of the food. This is to be expected when one considers the complex nature of foods. Water is an integral component of all organic materials. It is chemically and biochemically associated with the other foodstuff components, i.e. it is not present in a simply passive role. Moisture removal will also modify the cell structure thereby affecting the residual moisture content. By comparison, a mixture of an inert material, such as sand, and water can be dried in a predictable manner.

In the initial stage of drying of a food, the rate of moisture migration from the interior of the particle to the surface is sufficiently high to cover the surface in moisture. Under these circumstances the rate of drying of the particle is controlled by the rate of evaporation from the surface. As outlined above, this is controlled by the condition of the air adjacent to the surface. At this stage the food will dry as rapidly as the air can remove the moisture. This period of drying is known as the constant rate period. A critical moisture content will be reached when the moisture can no longer be drawn to the surface fast enough to maintain a completely wet surface. This can be attributed to physical and chemical changes occurring within the food. Once this critical moisture content has been reached, the rate of drying decreases. This second phase of the drying cycle is termed the falling rate period. Eventually the moisture content of the food will drop to a level where there is no driving force between the air and the surface, and drying will cease. The food is said to have reached its equilibrium moisture content.

It will be appreciated that, again, some broad generalisations have been made to give a general insight into drying mechanisms. In practical terms, two important points should be made which would benefit any drying extension work:

(i) for the initial drying of a commodity, the combination of high air velocity and moderate temperature will optimise the use of energy for drying;

(ii) in the latter stages of drying, low air flow combined with high air temperature will provide more rapid drying than a high air flow with a low temperature.

II. Solar radiation - the available energy

For practical purposes the sun's radiation to earth may be regarded as a beam of uniform intensity; just outside the earth's atmosphere that intensity is 1.3-1.4 kW/m^2 depending on the distance between the sun and the earth. This amount of radiation is in fact reduced to values of less than 1 kW/m^2 at the earth's surface by the presence of clouds, dust particles, and gases. The magnitude of the available radiation also depends on the location, time of year and time of day. The effect of absorption, mainly by ozone, water and carbon dioxide molecules, and scattering by dust particles, air molecules and water vapour is to both lessen the total amount of radiation per unit area and also to alter the proportion of the different wavelengths present in the radiation. The result of this is to decrease greatly the amount of ultra-violet radiation due to absorption by ozone, and also to decrease the amount of infra-red radiation reaching the earth's surface. The annual mean global irradiance is shown in figure 1.

Radiation scattered in the atmosphere is not entirely lost to the earth; about half of it reaches the earth's surface in a diffuse scattered form. Solar radiation can therefore be seen to consist of two components: diffuse and beam radiation. Each component has different characteristics which affect solar drying.

Diffuse radiation cannot be concentrated by means of focusing devices. On a cloudless day, with a clear atmosphere, about a fifth of the solar radiation available to a horizontal surface consists of scattered radiation coming from the whole sky. On an overcast day the proportion may reach 100 per cent. Humid coastal strips will experience high levels of diffuse radiation due to the moisture held in the air. Arid areas may also have higher levels of diffuse insolation than expected, due to dust in the atmosphere.

For the purpose of solar drying it might be thought that the positioning of the collector should be horizontal to optimise absorption of diffuse insolation. However to maximise the effect of beam or directional radiation, the collector surface should be tilted at right angles to the incident beam.

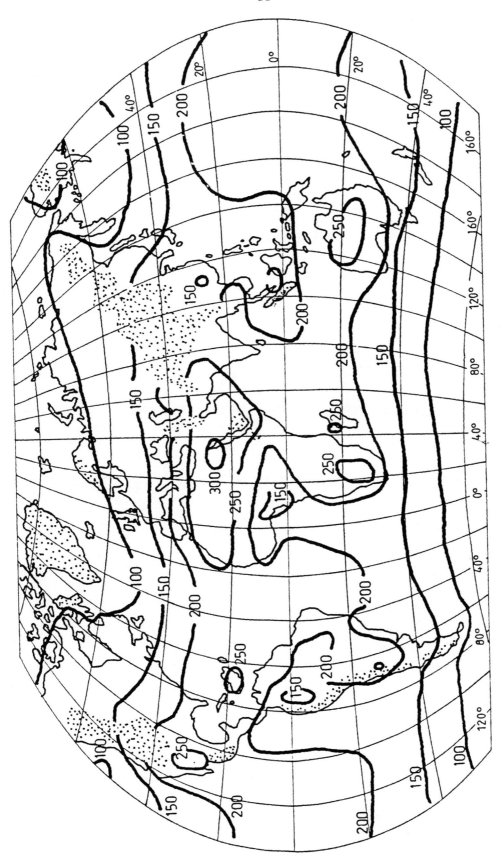

Figure 1

Annual mean global irradiance on a horizontal plane at the
surface of the earth (W/m^2 averaged over 24 hours

Source: Budyko (1958)

An inclined surface, moreover, may receive a significant proportion of its irradiance through reflection from the ground adjacent to it.

Beam radiation comes in a beam directly from the sun. Its presence can be easily recognised by its ability to cast shadows. The sharper the shadow, the greater the amount of direct radiation that is present.

While the extension worker need not concern him/herself with the theory of insolation mechanisms, he/she should be aware of the two main characteristics described. In practice, diffuse radiation will always be present and is useful for drying. However, the solar dryer should be orientated to make maximum use of the beam radiation component.

II.1 Direction of beam radiation

The relative movement of the sun and earth must also be taken into account in designing solar dryers and collectors.

The position of the sun in the sky is, for a given location, dependent upon the time of day and year. The daily movement of the sun - rising in the East in the morning to its highest point at mid-day and then setting in the West - is due to the rotation of the earth about its own axis.

The change in climate with the seasons is a result of the tilt of the earth's axis and its orbit about the sun. The period taken to complete one orbit is a year. The angle of tilt of the axis to the plane of the orbit is approximately 23.5°. The hemisphere (north or south) which is angled towards the sun at a particular time during the orbit around the sun will be receiving sunlight more directly and for a greater time each day than the other hemisphere.

The changes in the sun's position with time of year can be described from the point of view of an observer on the earth's surface. On June 22 the sun is at its most northerly point and appears directly overhead at mid-day on the Tropic of Cancer (23.5°N). As a result, the northern hemisphere receives a relatively large amount of sunlight at this time of year. As the year progresses the sun appears to move south giving longer periods of daylight in the southern hemisphere, but less in the northern hemisphere. On September 22 the sun is directly overhead at the equator and both hemispheres receive similar amounts of insolation. The sun continues to move south until December 22, when it is directly overhead at the Tropic of Capricorn (23.5°S). On this date, the northern hemisphere has its shortest period of daylight and the southern hemisphere its longest. Having reached its most southerly point, the sun then moves northwards, crossing the equator again on March 21, and is again overhead at the Tropic of Cancer on June 22 to complete its yearly cycle.

An example of a practical application of this information can be found in Kenya, a country located on the equator, where solar assisted coffee dryers are being put into operation. The dryers are built and operated at two different locations with different harvesting seasons. At one location the major crop is picked in November and December. From our knowledge of the movement of the sun, it is clear that at this time of year the sun is directly overhead near the Tropic of Capricorn. Thus, to maximise the level of insolation on the collector, the collector must face south. At the other location the main crop is picked in April and May when the sun is in the northern hemisphere. For this dryer the solar collector faces north, again to maximise the level of insolation upon it.

The important point to note here is that the solar collector is designed such that it is perpendicular to the sun at solar noon on the day selected as representing the peak of the harvest.

II.2 Solar collectors

Solar collectors are employed to gain useful heat energy from the sun's radiation. They are almost invariably used to heat either air or water. For the purpose of food drying, simple flat plate air-heating collectors can provide the desired temperature increase. These consist of an absorbing surface which heats up and warms the ambient air nearest to the surface. Clear covers may be placed above the absorber to reduce heat loss, and the collector unit may be insulated. Where a relatively high air flow is required a fan can be used to blow air through the collector. However, natural convection systems are widely used and may be more appropriate to the needs of the small-scale farmer.

There are therefore two stages in which the energy of the sun's radiation is transformed to thermal energy in the drying air. Firstly the radiation must be absorbed on the absorber surface, thus heating the absorber plate. This heat is then transferred to the air by contact between air and the absorber plate.

II.3 Absorber performance

Factors affecting the amount of energy absorbed by the absorber plate are:

1. **The level of insolation.** Clearly, the higher the insolation, the greater the energy absorbed. It should always be borne in mind that insolation levels vary considerably from place to place and at different times of the year;

2. **The angle between incident insolation and the absorber plate surface.** As previously discussed, the slope of the collector should be determined by considering the position of the sun -the declination angle - at the peak harvest time. The movement of the sun over the year is illustrated graphically in figure 2. To determine the correct slope for the collector the angle of declination at peak harvest time should be added or subtracted - depending on the harvest season - from the latitude of the dryer. In all cases, the collector should be positioned parallel to the equator. An example of this calculation is shown in table 1. To discover the latitude for a given location, one should consult an atlas. In most cases, the absolute angle used is not critical and if necessary a near estimate will usually suffice. It should be noted that the angle subtended by the transparent cover is not important;

3. **The absorptivity of the absorber surface.** The greater the absorptivity of the absorber surface, the higher the proportion of incident radiation that will be absorbed; in collectors where the absorbing medium is of a porous nature, it is also desirable that the absorber give little resistance to the passage of air. One of the more commonly used absorbers is black-painted metal sheets; frequently these are corrugated galvanised iron. This has the advantage in most places of being readily available, relatively cheap, and easy to use. Other suitable absorbers are black plastic sheets, painted rocks, ash and charcoal. These materials are suitable because they exhibit the following properties:

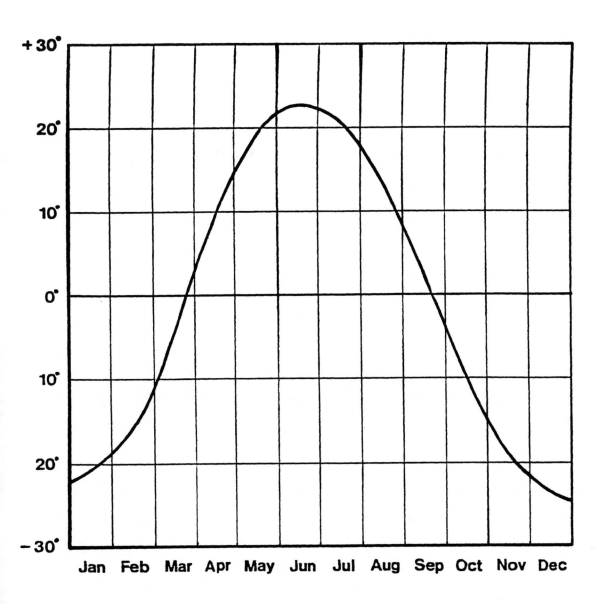

Figure 2

Variation of declination angle, with time of year

Location	Latitude	Time of peak harvest	Declination	Slope	Facing
Khartoum	+ 15	Mid April	+ 10	+ 5	South
		Mid October	- 10	+ 25	South
Lusaka	- 15	Mid April	+ 10	- 25	North
		Mid October	- 10	- 5	North

Table 1

Examples of the determination of slope angle

(i)   high absorptivity of incident radiation;

(ii)   low emissivity;

(iii)   good thermal conductivity in cases where the air flow is below the absorber;

(iv)   stability at the temperatures encountered during operation and under stagnation conditions;

(v)   durability;

(vi)   low cost;

(vii)   low weight per unit area.

These properties should be explained in suitable terms to an extension worker.

4.   <u>The transmissivity of the cover material.</u>   The ideal cover will permit the passage of sunlight, but not the longer infra-red (heat) radiations which are emitted by the absorber surface.   For most food drying operations, where temperature rises of up to $35^{o}C$ are sufficient, single cover collectors are adequate.

Glass is traditionally used as glazing.   It has good transmissivity for visible radiation and is virtually opaque to infra-red radiation.   Glass is also stable at the temperatures encountered and durable.   The disadvantages of glass are its low shatter resistance, high cost, and a large weight which increases the cost of the supporting structure.

The use of plastic sheets has in the past been limited by the poor weatherability and stability of plastics in the conditions found in solar collectors.   However, plastics have recently been developed which overcome these problems.   Examples of these are polyvinylfluoride films (PVF) such as Tedlar; fibre-reinforced polyester (Kalwall, Sun Lite Premium); and acrylic and polycarbonate sheets which are either intrinsically stable to UV radiation or have been made stable by use of additives.   Plastic covers weigh as little as 10 per cent of the weight of glass covers, (for example, a 1 m x 1 m PVF sheet 1 mm thick weighs 3 kg compared to over 30 kg for a 5 mm glass sheet of the same area).   Other materials such as acrylic and polycarbonate sheet of the same thickness as glass sheet weigh about half as much as glass for the same area.   Plastics also have the advantage of easier handling and installation.   On many locations, the choice of which plastic to use will be limited to what is locally available. Heavy-duty polyethylene film may be manufactured locally by a packaging materials factory. Polyethylene has comparatively poor cover properties, since it transmits 77 per cent of the infra-red radiation generated and can become opaque with exposure to sunlight. It is also more likely to tear than some other plastics. However, it has the advantages of cheapness and relative ease of availability. One interesting source of transparent, heavy duty plastic is the material used to protect car seat covers. Such material can be found in many developing countries.

When considering cover materials the important points to consider are:

(i)   high transmissivity on the visible range of the spectrum;

(ii)   low transmissivity of infra-red radiation;

(iii)   stability at the operating temperature. All materials used must be able to withstand the temperatures attained under stagnation conditions, i.e. on a hot sunny day when no air is flowing through the collector;

(iv)   durability or weatherability;

(v)   strength and resistance to breakage;

(vi)   low cost;

(vii)   low weight per unit area.

In most cases, plastic covers will be most suitable. The reason for using these should be explained in appropriate terms to the extension worker.

III. Classification of solar dryers

Solar dryers can be classified using the following characteristics:

1.   Whether or not the drying commodity is exposed directly to insolation;

2.   The means of air flow through the dryer;

3.   The temperature of the air circulated to the drying chamber.

III.1 Exposure to insolation

Based upon this criterion, solar dryers can be termed either direct or indirect. Direct dryers are those in which the crop is exposed to the sun, and indirect dryers, those in which the crop is placed in an enclosed drying chamber and thereby shielded from insolation. In direct dryers, heat is transferred to the drying crop by convection and radiation, and therefore the rate of drying can be greater than for indirect dryers. For some commodities, (e.g. certain varieties of grapes and dates), exposure to sunlight is considered essential for the required colour development in the dried product. For arabica coffee in Kenya a period of exposure to sunlight is thought necessary for the development of full flavour in the roasted bean. On the other hand, with some fruits the ascorbic acid (vitamin C) content of the dried product is considerably reduced by exposure to sunlight. Colour retention in some highly pigmented commodities, such as green legumes, can also be adversely affected if they are exposed directly to sunlight.

III.2 Means of air flow

There are two possible types of air flow, natural convection and forced convection. The former is reliant upon thermally-induced density gradients for the flow of air through the dryer, whereas for forced convection dryers, the air flow is dependent upon pressure differentials generated by a fan. A fan is obviously capable of providing a much greater air flow and is therefore suitable, if not essential, for dryers with large throughputs. Another advantage of forced convection dryers is that the air flow is independent of ambient climatic conditions and is easily and accurately controllable for most applications. Forced convection is essential for the drying of deep beds of grain wherein the relatively high pressure-drops through the depth of grain would preclude the use of natural convection.

However, one fundamental disadvantage of forced convection dryers lies in their requirement of a source of motive power for the operation of the fan. The capital cost of equipment necessary to provide forced convection is high compared with the minimal costs of natural convection dryers. It may prove necessary to import the fans and the generator required to drive them. In addition the running costs of forced convection drying (power, mechanical maintenance, and repair) are high. These are disadvantages which indicate that forced convection solar dryers are probably not appropriate for use in rural areas by small-scale farmers.

III.3 Circulated air temperature

The air entering the drying chamber of a solar dryer can either be at the ambient temperature or at some higher temperature; the elevation in temperature of the air being achieved by its passage through a solar collector prior to the drying chamber. Dryers that employ a separate solar collector and drying chamber are usually more efficient, as both units can be designed for optimum efficiency of their respective functions, whereas dryers in which the collector and the drying chamber are combined are invariably less efficient. However, a dryer with a separate collector and drying chamber can be a relatively elaborate structure while the combined collector and drying chamber is much simpler and more compact. In practical terms, the advantage of simplicity may over-ride other design considerations, and the combined unit would certainly be the easiest type of solar dryer to introduce in an area unfamiliar with complex construction techniques.

Based on the above classification, three major groups of solar dryer development have been identified.

1. Group one: <u>Direct</u> dryers employing <u>natural</u> convection with a combined solar collector and drying chamber.

2. Group two: <u>Direct</u> dryers employing <u>natural</u> convection with <u>separate</u> collector and drying chamber.

3. Group three: <u>Indirect</u> dryers employing <u>forced</u> convection with <u>separate</u> collector and drying chamber.

IV. Group one dryers

IV.1 Type one - cabinet dryers

Within this subclassification several variations have been developed. Essentially, the basic design (figure 3) consists of a rectangular container, preferably insulated, and covered with a roof of glass or clear plastic. There are holes in the base and the upper parts of the cabinet and rear panels. The interior of the cabinet is blackened to act as a solar absorber. Perforated drying trays are positioned within the cabinet. Access to the trays is through doors which form the bottom part of the rear of the cabinet.

The operating principle of the cabinet dryers is that insolation passes through the clear cover and is absorbed on the blackened interior surfaces

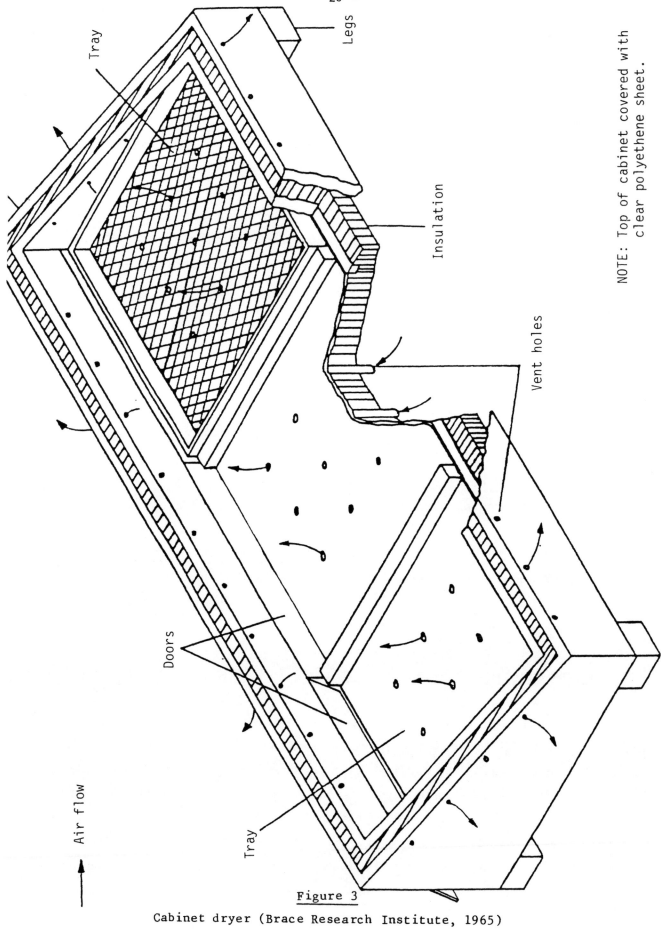

<u>Figure 3</u>

Cabinet dryer (Brace Research Institute, 1965)

which are thereby heated and which subsequently warm the air within the cabinet. The warmed air rises by natural convection and passes up through the drying trays and out of the cabinet via the upper holes, whilst fresh air enters through the holes in the base.

The construction of a cabinet dryer is demonstrated in plates 2.1 - 2.4.

It is recommended that the length of the cabinet be three times its width to minimise the shading effect of the sides. In any situation the roof should be angled sufficiently to allow water to run off in rainy periods. For portable models the cabinet may be constructed of wood, board or metal for the more sophisticated units, and material such as wicker or basket work for more rudimentary models. For permanent (and larger) structures mud, brick, stone, or even concrete could be used. The insulation for the base and sides can be wood shavings, sawdust, bagasse, coconut fibre or dried grass or leaves. It is recommended that the insulating layer be at least 50 mm thick for maximum efficiency and effectively sealed in place to prevent the ingress of moisture and insects and the like. Where insect infestation is troublesome, all air holes in in the cabinet can be covered with gauze or mosquito netting if these are cheaply available. The drying trays can be constructed of plastic mesh or netting, or even of wicker and basket work, but preferably not of metal since this may adversely react with the juices from fruit or vegetable slices. The temperature within the cabinet is regulated by the inlet or outlet holes and by the degree of opening of the access doors.

A considerable amount of investigatory work has been carried out in Jamaica on the development of cabinet dryers for a large number of tropical crops. In this case dryers were constructed from clay bricks, wattle and daub or locally produced compressed blocks made of earth and cement. Plastic sheet was used as a cover material. A length of bamboo placed at the bottom of the cabinet with regularly spaced holes was used as the air inlet pipe. Air outlet ports were made by leaving gaps in the top layer of bricks. Charcoal fines mixed with clay were used as an alternative to paint for blackening the interior of the cabinet.

One method of extending the drying time of a solar cabinet in periods of inclement weather or at the end of the day is the inclusion of a heat store. This can be brought about by placing a layer of dark coloured or black painted stones in the bottom of the dryer. However it should be appreciated that

Construction of a cabinet dryer

Plate 2.1

The outer box, in this example, is made from seasoned timber

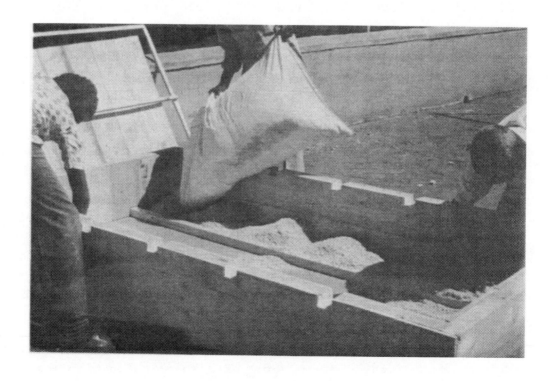

Plate 2.2

A lining of sawdust provides insulation between the two layers

Plate 2.3

Applying the inner floor. Note that the batten positions should
be marked to allow air holes to be drilled through both thicknesses

Plate 2.4

The completed dryer with trays and lid awaiting
painting and the plastic cover

this practice will increase the thermal inertia of the system, which means that the dryer will take longer to heat up in the beginning. In some cases this controlling action may be desirable if it prevents the cabinet from overheating and cooking the product.

It is worth mentioning the conversion of a cabinet to indirect use within this subgroup. A black cover can be placed under the clear cover to reduce loss of colour and nutrients as a result of direct exposure to sunlight. This modification would be suitable for drying green leafy vegetables.

Cabinet dryers have the most widespread use and as such have been tested with a range of commodities including fish, fruit, vegetables, root crops, and oilseeds.

IV.2 Type two - tent dryer

The second popular type within this group of dryers was originally developed for use with fish. The dryer (figure 4) consists of a ridge tent framework, covered with clear plastic sheet on the ends and on the side facing the sun, and black plastic sheet on the side in the shade and on the ground within the tent. A drying rack is positioned centrally along the full length of the tent. The plastic sheet at one end is arranged so as to allow access to the rack as required, but otherwise is fastened shut. The bottom edge of the side of clear plastic is rolled around a bamboo pole which, when raised or lowered, forms a method of controlling the air flow through and the temperature within the tent. Holes in the apex of both ends of the tent permit the venting of the exhaust air.

The advantages of the tent dryer are its simplicity of construction and operation, and its low cost. However it has the disadvantages of a high plastic content in relation to the other construction materials. It is lightweight and susceptible to damage in windy conditions.

V. Group two dryers

Development of this type of dryer has been largely carried out at the Asian Institute of Technology in Thailand. The dryer (figure 5) consists of two separate units, the collector and the drying chamber. The solar collector uses a layer of burnt rice husks or black plastic sheet for the

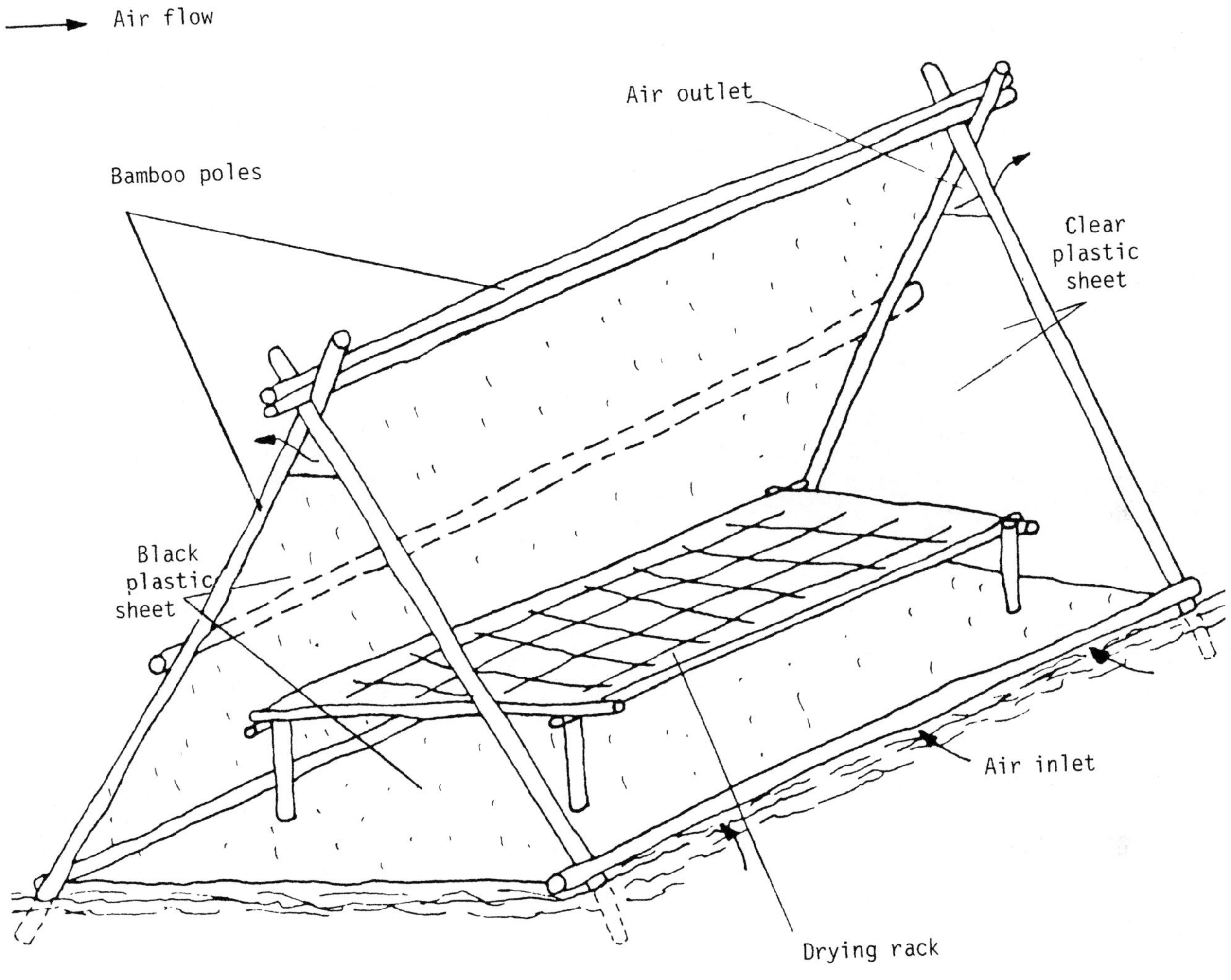

Air flow

Air outlet

Bamboo poles

Clear plastic sheet

Black plastic sheet

Air inlet

Drying rack

Figure 4

Solar tent (Doe, 1979)

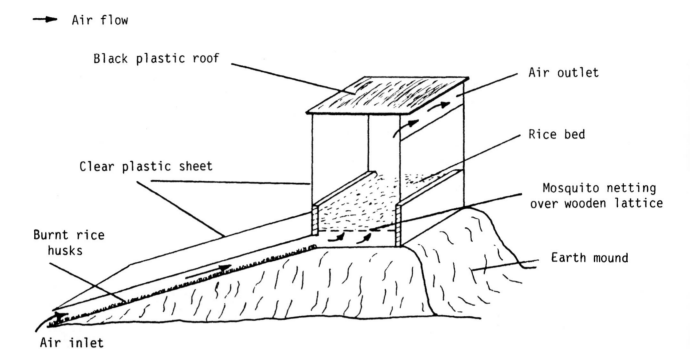

Figure 5
Solar paddy dryer
(Exell and Kornsackoo, 1978)

absorber surface. This is covered by clear plastic sheet on an inclined bamboo framework. The collector warms the ambient air which is then passed to the base of the drying chamber. The drying chamber can be a simple shallow wooden box with the base made of either perforated metal, plastic mesh, or bamboo matting. The dryer is loaded through removable panels at the back of the drying chamber. The foodstuff to be dried is placed on top of the perforated base. Since this type of dryer was designed to dry rice, the use of a metal base is quite acceptable. The heated air passes through the drying chamber and is exhausted through the top of the unit. At this point a chimney can be placed on top of the drying chamber to provide a column of warm air to increase the draught and hence the flow of air through the dryer. At its simplest, the chimney can be a bamboo frame covered with black plastic sheet acting as an absorber. The air inside the chimney will heat up, rise and thereby pull air from the collector through the dryer.

The drying chamber can also be constructed (figure 6) using a wooden frame covered in plastic. Black plastic, similar to wood painted black, will increase heat absorption in the drying chamber. Where the sides of the drying chamber are made from clear plastic, overall insolation will be increased. However, where rice is being dried using this arrangement, it is recommended that the roof be made from black plastic to avoid overheating the uppermost layers of paddy in order to reduce the cracking. The construction of a modified paddy dryer is shown in plates 2.5 - 2.8.

The reader will appreciate that the chimney type dryer is a more sophisticated device than the cabinet dryers previously described. As such its construction will be more demanding on the artisanal skills available, and the principles of its operation more difficult to comprehend by the end user.

Other potential shortcomings of this type of dryer are its relatively high profile, which can pose stability problems in windy conditions, and the fact that extensive areas of the dryer surface are composed of plastic sheet, the replacement of which can be a relatively expensive undertaking.

VI. Group three dryers

This group of dryers can be considered as devices consisting of a solar collector (usually of a flat plate type) and a drying chamber with a fan moving air from the collector to the drying chamber. Dryers within this

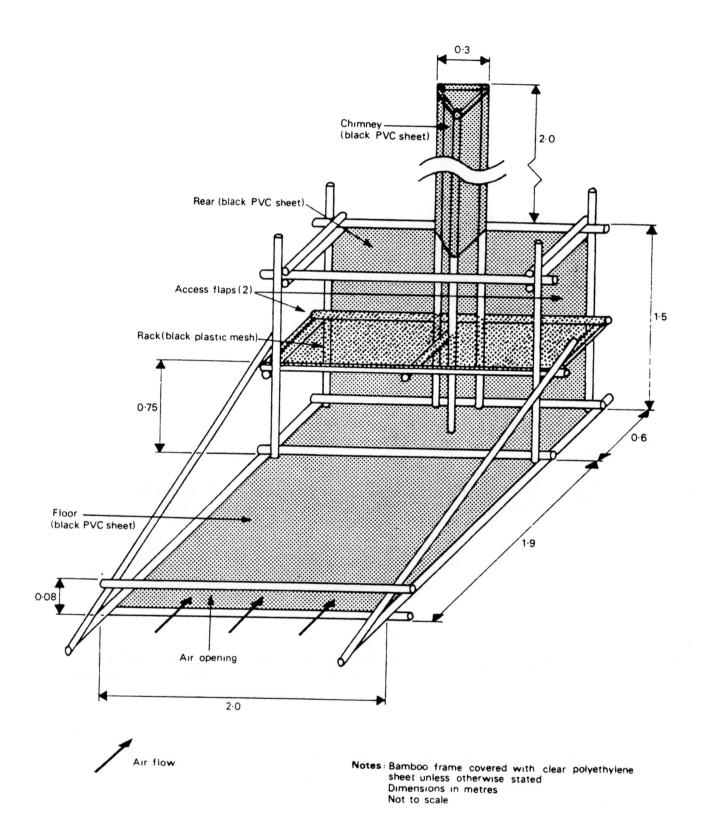

0·3

2·0

Chimney
(black PVC sheet)

Rear (black PVC sheet)

1·5

Access flaps (2)

Rack (black plastic mesh)

0·75

0·6

Floor
(black PVC sheet)

1·9

0·08

Air opening

2·0

Air flow

Notes: Bamboo frame covered with clear polyethylene
sheet unless otherwise stated
Dimensions in metres
Not to scale

Figure 6

Group two dryer

Construction of a modified solar paddy dryer

Plate 2.5

The basic framework

Plate 2.6

Instead of black plastic, thin painted plywood was used
for additional strength

<u>Plate 2.7</u>

Careful application of the polyethylene cover

<u>Plate 2.8</u>

The completed dryer

group have been used primarily for the drying of grain, particularly in the cool, but dry autumns of the grain-growing areas in North America, but also in the warm, but humid climate of Asia. Compared with the more basic natural convection systems, this type of dryer is more capital-intensive due to costs attributable to the fan, and tends to be more suitable for larger outputs. Because of the high construction, maintenance, and running costs, such systems are usually carefully engineered to give the most efficient operating conditions. The construction materials required are of a higher specification than found in many rural locations, and the construction and operating tolerances are controlled to ensure optimum output from the dryer. The applicability of this type of dryer in the rural sector of developing countries is inevitably very limited.

VII. Hybrid dryers

This term is used to denote those dryers in which another form of heating the drying air is used in conjunction with solar heating. Such a system can be used in two ways. Firstly, solar heating can be the principal source of energy during sunny daylight hours with additional heat, supplied by electricity, solid fuel, etc. being used during inclement weather or, in some cases, at night to maintain continuous drying. Secondly, conventional energy sources are used as the main means of heating the drying air, and solar energy is used as a supplement to reduce fuel costs. The latter system has been extensively researched, particularly in the grain-growing areas of the United States. In some instances, storage silos have been modified by the incorporation of solar collectors into the walls and/or roof.

CHAPTER 3

DEVELOPMENT AND EXTENSION OF SOLAR CROP DRYING TECHNOLOGIES

yer selection and design

Having established that solar drying of a particular commodity may be
cially viable and having gained an appreciation of the different types of
dryers and their respective advantages and disadvantages, the next step
e selection and design of the most suitable type of dryer.

In order to be able to choose the appropriate size of solar dryer it is
tant to be fully aware of the quantity of crop to be dried over the
, the pattern of harvesting, the relationship between batch size and
g time, and the effect of varying drying time on product quality.   It is
ese that the dryer size, the number of dryers required, the batch size,
xpected drying time can be determined.   It should be noted that a
tion in drying time in itself may not necessarily benefit the user
cially.   For example, a farmer who picks a crop once a week has little
in by reducing the drying time for the crop from four days to three days,
s there is a recognisable improvement in quality from more rapid drying.

It may not always be necessary or desirable that a dryer be designed to
e all of a farmer's or cooperative's production.   This is particularly
the case of simple natural convection dryers for use at the rural
.   The dryer design should be based on what the true drying load is
ated to be and not on the total crop harvested.

The dryer should be easily constructed from materials readily and cheaply
able to the end-user using locally available skills.

When considering the size of the dryer it should be borne in mind that
r dryers may require stronger and, hence, more expensive construction
ials than smaller dryers.   The dimensions and the strength of the plastic
available should be taken into consideration.   For a large unit, the
may be made up from individual, smaller plastic sheets.   Joining one
to another can be difficult and may give rise to problems such as tears

and air leaks. For these reasons it may be more appropriate, in the case of simple natural convection dryers, to build and operate a number of small units. Multiplicity allows diversity, for more than one crop can be dried at a time. A further advantage is that if one dryer is out of function due to damage, drying can still continue, at reduced capacity, using the other dryers.

On the other hand, more sophisticated dryers, such as forced convection solar dryers, benefit from economies of scale due to the investment tied up in the fan and source of motive power. Generally speaking, one large dryer will be more cost-effective than two smaller units. However it should be taken into consideration that an oversized unit will be operating at less than full capacity negating any cost advantage. The drying area required will depend on local conditions and the commodity, but as a rule of thumb, one square metre will accomodate approximately 10 kg of fresh produce.

I.1 Construction methods and materials

Construction methods and available materials may vary considerably from location to location. It is not within the scope of this manual to discuss individual, local circumstances. Some general guidelines regarding factors which must be considered can, however, be given:

(i)   dimensions of standard materials. Where possible, designs should take account of the sizes of material locally available. For example, it would be poor design to specify the width of a corrugated iron collector as 1.1 m if the standard width of corrugated iron sheet is 1 m. Before finalising a design, the commercial availability of materials must be ascertained.

(ii) use of rural materials. The cost of building a solar dryer can be minimised if the producer is able to use wood cut straight from the forest rather than prepared timber. Careful design in the development stage of a dryer can often facilitate the use of cheaper materials. Difficulties caused by these materials are in joining pieces of the structure, in sealing the structures against air leaks, and in attaching the plastic sheet to the (wooden) frame. There is obvious scope for designs which use prepared timber at strategic points and unprepared at others. Where the use of wood is necessary, remember to take environmental factors into consideration.   For example, determine what the effect of flash flooding or termites might be and take the appropriate preventive action.

(iii)   use of plastic sheet.   For many solar dryers, the clear plastic sheet used is the major capital cost to the farmer; therefore, the type of plastic chosen is important.   A choice must be made between a relatively cheap plastic such as ordinary polyethylene which will last, at best, for one season due to photo-degradation and wear and tear; and a more expensive, better quality plastic less prone to photo-degradation; or even glass or a rigid plastic.

Attaching plastic sheet to the framework structure, so as to minimise the likelihood of the plastic being torn is, perhaps, the most difficult part of building a dryer.   Listed below are some general points which should be followed to prolong the useful lifetime of plastic sheet on a solar dryer:

(i)   when attaching plastic sheet to the framework, care should be taken not to stretch the plastic at the points of attachment, but the plastic should not be so loose that it will flap about in the wind.

(ii)   rather than merely stapling or nailing the plastic directly to the framework, it is preferable to sandwich the plastic between the framework and a batten.   This may not be practical when unprepared wood or other materials are being used;

(iii)   no sharp edges should come in contact with the plastic sheet since these will initiate tears;

(iv)   fold over the plastic at the point of attachment to the frame, so that there are two or more layers of plastic. This will help prevent tears;

(v)   when fixing the sheet over the framework, sags and hollows in which water can collect should be avoided wherever possible;

(vi)   the dryer should be handled as carefully and as seldom as possible during operation and when not in use.

The comparative merits of natural convection and forced convection systems have already been described. It has been indicated that simple natural convection dryers will probably have the more widespread application in the rural sectors of developing countries. The information contained in the manual should be sufficient to enable simple dryers to be constructed and operated for a range of commodities. Where the reader feels that a forced convection dryer would be appropriate, he/she is advised to obtain more detailed information before attempting the construction. The reader will need to know about fan delivery mechanisms to ensure the correct air flow rates through the dryer. Efficiencies at various stages of the drying process will have to be measured, including collector efficiency and overall drying efficiency. Furthermore, the performance of the dryer in its working location will have to be routinely monitored to ensure continuous performance at maximum efficiency. This implies that a continuing input to such a development would be necessary by an individual well versed in solar drying theory and practice. The reader may find that such skills are available to him/her locally at national institutes. There are, to the author's knowledge, solar research and development projects where the required specialist knowledge is available in many countries in South and South East Asia and in most of the African Commonwealth nations. The reader is also referred to the list of references at the back of this manual for further information.

To summarise, the technologist will need to obtain a range of technical and socio-economic information to allow the selection of the most appropriate dryer. The technologist should then present a complete package to the extension worker. This should include a working dryer with information in a suitable form on its construction and use. Details of drying techniques can be found in the individual commodity chapters together with suggestions on how to put the information across.

1. Technical criteria

   The following design factors must be established:
   - the throughput of the dryer over the productive season;
   - the size of batch to be dried;
   - the drying period(s) under stated conditions;
   - the initial and desired final moisture content of the commodity (if known);

- the drying characteristics of the commodity, such as maximum drying temperature, effect of sunlight upon the product quality, etc.;
- climatic conditions during the drying season, i.e. insolation intensity and duration; air temperature and humidity; windspeed; (such data may be available from local meteorological stations);
- availability and reliability of electrical power;
- the availability, quality, durability and price of potential construction materials such as:

    glazing materials: glass, plastic sheet or film;

    wood (prepared or unprepared),

    nails, screws, bolts etc.;

    metal sheet, flat or corrugated angle iron;

    bricks (burnt or mud), concrete blocks, stones, cement, sand etc.;

    roofing thatch;

    metal mesh, wire netting etc.;

    mosquito netting, muslin etc.;

    bamboo or fibre weave;

    black paint, other blackening materials;

    insulation material; sawdust etc.;

    fossil fuels (to power engines to drive fans if electricity is not available);

- the type of labour available to build and operate the dryer;
- the availability of clean water at the site for preparation of the commodity prior to drying.

In any one situation there may well be other technical factors that need to be considered.

2. Socio-economic criteria

From the initial considerations, estimates of the capital costs of the dryer, the price of the commodity to be dried, and the likely selling price of the dried product will have been made. Other questions that need to be considered are the following:

- who will own the dryer?
- is the dryer to be constructed by the end-user (with or without advice from extension agencies), local contractors, or other organisations?
- who will operate and maintain it?

- how can the drying operation be incorporated into current practices?
- are sources of finance from local authorities or extension agencies available, etc.?

Obviously there are many other socio-economic factors, particularly those of a local nature, which must be taken into account. It cannot be stressed too highly that if such factors are not taken into account and evaluated, then there is every chance that an inappropriate dryer design may result. Equal emphasis must be placed on both technical and socio-economic factors.

Summary

(1) Situations where solar dryers may be useful:

- where the cost of conventional energy is prohibitive and/or the supply is erratic, to supplement existing artifical drying systems and reduce fuel costs;
- where land is in short supply or expensive;
- where the quality of existing sun dried products can be improved upon;
- where labour is in short supply;
- where there is plenty of sunshine, but high humidity.

(2) Situations where solar dryers may not be useful:

- where conventional energy sources are abundant and cheap;
- where large amounts of combustible by-products or waste materials are freely available;
- where there is insufficient sunshine;
- where there is plenty of sunshine and arid conditions (sun drying may suffice);
- where the quality of sun dried products already made cannot be improved upon;
- where local operators are insufficiently trained;
- where the ramifications of introducing a solar dryer have not been completely thought out.

II. Extension techniques

The normal route for the introduction of a new technology such as solar drying is research, development, and then extension. The three steps are inter-related. A technology which has been tried and tested and found to be good is the one which is most likely to be widely adopted irrespective of whether there is large extension effort. It can be argued that the greater the effort which is spent ensuring that the correct technology and level of application have been identified, the easier will be the job of the extension worker. The converse is also the case in that no amount of well-implemented extension work will result in the successful introduction of an inappropriate or ill-conceived technology. Extension cannot therefore be discussed in isolation from research and development.

In this section some of the problems likely to be encountered in efforts to promote the solar drying technology to the rural population will be considered. Suggestions will also be made on methods of overcoming these problems. A note of caution needs to be added here: the examples cited may be pertinent only to dissemination of engineering technology to agriculture, and are not necessarily typical of general extension practices.

II.1 Prerequisites to successful dissemination of engineering technologies in agriculture

Many extension experts agree that engineering technologies are among the most difficult technologies to disseminate to farmers, particularly to smallholder farmers. Engineering technologies include improved and modern cultivation implements and machines, as well as post-harvest processing techniques such as solar drying. By comparison, the extension of agricultural improvements may be relatively simple. An example often quoted is the successful adoption of hybrid maize seeds by small-scale farmers in East Africa. By contrast, modern farming tools such as oxen-powered cultivators and tractors have not been adopted to any great extent by the same sector, where about 80 per cent of the cultivated land is still being worked with hand tools.

In many cases engineering technologies have the objective of optimising the use of labour as well as eliminating drudgery. Several reasons have been given for the limited adoption of engineering technologies in agriculture:

(i) engineers have tended to design expensive and complicated equipment or implements which are beyond the financial and technological capabilities of the farmers and even of the extension workers;

(ii) these technologies have often aimed at optimising labour use and hence they are perceived as leading to increased unemployment, especially in rural areas;

(iii) these technologies have relied on imported inputs which are difficult to obtain locally;

(iv) research and development by engineers have often been done in laboratories without any regard to the situation in the field.

It is often forgotten that engineering technologies demand that the user have a level of technical knowledge higher than that required for the adoption of agricultural improvements. The introduction of a new hybrid maize may result from years of sophisticated intense research, but what is eventually offered to the farmer is seeds which he/she has to plant in a particular way. It is not necessary that the farmer be aware of the genetic changes in the hybrid stock. However, in order to appreciate the advantages of a new implement such as a solar dryer, a higher level of technical appreciation is required. It is therefore of great importance that the front-line extension worker be technically well-informed.

For successful extension of solar drying technologies there is therefore a need for:

(i) proper evaluation and experimentation before extension. The whole system in which the farmer and extension agent are working should be understood, and the technology should be developed to suit the local scene rather than expecting that the local scene will adapt to the technology;

(ii) the front-line extension worker should be well trained and conversant with the new technology. If the extension agent does not understand the new technology it is unlikely that he/she will extend it effectively, and likewise, it is unlikely that the farmer will understand and adopt it;

(iii) the technology must aim at solving an issue which both the farmer and extension agent see as a problem and which the farmer regards as a priority problem among his many others. There is no point, for example, of concentrating on a solar maize dryer when the farmer's principal problem is organising land preparation, planting and weeding to exploit the short rainy season common in his locality. He will certainly not invest in a dryer if he is not even sure of harvesting the maize to be dried;

(iv)   the technology must aim at solving the problem by optimising the use of labour at a reduced level of human energy expenditure, i.e. the technology must not increase drudgery;

(v)   the technology will have most chance of success if the need has been recognised by the main decision maker of the farm or cooperative organisation.   The decision maker will be more enthusiastic if the advantages of the improved technique benefit him directly;

(vi)   there must be a financial or other obvious incentive to adopt the new technology;

(vii) the technology must be within the financial resources of the farmer and must be easily and inexpensively repaired if there is any breakdown.

The important thing to note is that the solar drying will not be operated in isolation from the environment; all the external and internal factors which affect the system need to be considered.

II.2 Extension work outlets

The solar drying technologist must decide which extension service, if there is more than one, will be the most suitable to be responsible for carrying out extension work in improving sun and solar drying technologies. There may be several different extension networks in the country each having its own priorities and areas of operation.

Examples of the possibilities, which are by no means representative of any one country, are discussed below.

1.   front-line extension workers; These are normally government employees working for the various ministries responsible for agriculture,

fisheries, livestock etc. It is likely that these employees will be trained to certificate or diploma level at a national college. In some cases the extension staff will have received all their training on the job and will hold their current positions due to long working experience in the sector. Extension workers at this level have the advantage of local knowledge and close working relations with the rural population. However, the front-line extension worker's knowledge of engineering principles is usually quite low. At a management level above this, the extension coordinator for a region or district, while not actively involved in the field, may have some specialist knowledge or have other specialists reporting to him who would be capable of implementing the proposed programme. The responsibility for translating the operating requirements accurately into locally comprehensible terms may however remain with the reader.

2. crop authority extension agents; In some countries extension networks may be developed for a specific commodity, particularly where the national importance of a crop has been recognised. Foodstuffs covered in this way may include coffee and fish where production may be controlled by a quasi-autonomous produce board within the appropriate ministry. These specialised agencies can be effectively used to extend new technologies such as solar drying since they will be aware of the problems specific to the crop and may also be aware of the techniques required to overcome these problems.

3. extension departments in academic institutes; Many research institutes and university faculties have their own extension departments which carry out some extension work. They are normally staffed by highly trained extension experts, usually educated to at least first degree level. One limitation is that, in some instances, the department's primary consideration is the theory of different extension techniques and socio-economic studies of extension impact rather than practical extension work.

4. rural development agencies; These bodies are often set up by ministries responsible for education or social welfare. They often have the brief of improving the lives of the rural people by introducing elementary health care, nutritional education, etc. Project areas include small-scale cultivation and simple preservation techniques. Extension

workers in these agencies might prove suitable contacts to promote the use of simple drying techniques. Again it should be emphasised that the information should be presented in a suitable form.

5. appropriate technology or rural technology centres; In recent years many appropriate technology centres have been established with differing mandates but essentially for the development and adaptation of technologies to local conditions. These centres may prove to be suitable vehicles for the transfer of solar drying technology.

6. commercial companies; Where there is an obvious need for a particular item of equipment, commercial companies can purchase the patent of a particular technology or else manufacture it under licence. The marketing departments of these companies then carry out the extension work. Since the companies are profit motivated, the initial evaluation of the process tends to be thorough. Once a commercial organisation enters into such an enterprise it can be assumed that the technology stands a good chance of being extensively adopted.

In addition to the above, there are other organisations such as religious organisations, women's groups, local cultural groups and schools which may be used to popularise the technology and lead to its wider adoption. The route which is most effective in propagation and hence widespread adoption of the technology is best determined by careful examination of the local situation.

II.3 Extension techniques

In addition to the route through which the technology reaches the farmer or target group, there are also different methods through which the technology can be demonstrated to farmers. Many of these techniques have been developed for agricultural technology - e.g. crop husbandry, livestock husbandry etc. and how effective they may be for engineering technologies is difficult to ascertain. The main techniques are:

(i) demonstration at market places/meeting places; In this case the extension officer takes the technology to a place where the people in that particular area gather in large numbers. This may be at a market place. He/she then exhibits the solar dryer (or any other technology) to the people providing as much information as possible;

This technique is more appropriate to the introduction of simple improvements such as a hybrid seed variety or small portable implements, where either the risk to the farmer is low or the improvement is obvious. It is preferable to have a permanent demonstration site for solar dryers where continuing drying practices can be demonstrated. Since drying takes several days to complete, it is not practical to demonstrate it in the market place. However, poster-displays in the local language and samples of the dried foods could be exhibited. Interested farmers could then be taken to the pilot-plant demonstration unit at the local extension office or rural technology centre;

(ii) progressive farmer technique; In this case a progressive farmer in a particular locality is selected and encouraged to use the new technology (either on loan or at a subsidised price) and if he/she is satisfied with its performance he/she can, through the demonstration effect, encourage other farmers to follow his example. These progressive farmers are usually the more educated and richer members of the community and quite influential. Due to their financial position, they are better able to take risks, which may not be the case for the less progressive and poorer farmers. However if a technology has been well evaluated and the benefits can clearly be seen, then the chances of wide-scale adoption through this extension technique are great;

(iii) focus and concentrate technique; In this case farmers are selected using criteria established by the extension officer according to the technology which is being extended, and they are given the technology (usually free of charge). The extension officers focus all their efforts on these farmers until the technology is adopted. If it is adopted then neighbouring farmers are likely to copy it. It is an expensive method of extension and in many cases the selection criteria result in the selection of only the progressive and influential farmers. Another disadvantage is that those farmers who are supposed to copy the technology later on will also expect a free initial input;

(iv) training and visit technique; This technique involves grouping farmers in groups of say 10-15 people with a group leader. These farmers then meet in a specified place and are given training on the new technology. This is followed by programmed visits to the farmers at specified intervals throughout the season when further training is given. It is an expensive method and it requires significant input in trained manpower resources as well as transport facilities.

Any of the above four techniques could be used to propagate the use of solar crop dryers. There may well be other methods which are more suitable. Again, the method to be selected will depend on local conditions as well as resources available to the extension officer.

One major problem to overcome is the cost of the dryer to the small-scale farmer. Even if the interested farmer could test the unit before buying it to satisfy him/herself of its usefulness, the purchase of a dryer or the construction costs involved still represent a major financial outlay for the poorest in the rural sector. This problem can be alleviated by encouraging co-operative ownership of the equipment where this is possible. A pricing mechanism based on the quality of the dried produce will also help to recoup the costs.

CHAPTER 4

FISH DRYING

Unless fish are preserved or processed in some way to retard spoilage
they will start to decay within a few hours of being caught at the high
ambient temperatures in tropical countries. Spoilage is caused by the action
of enzymes (autolysis) and bacteria in the fish, and also by chemical
oxidation of the fat which causes rancidity.

Salting and drying, used on their own or in conjunction with each other,
are traditional methods of preserving fish which have been used for
centuries. Dried salted products are still very popular in parts of Africa,
South and South East Asia, and Latin America. Reducing the moisture content
of fresh fish by drying to around 25 per cent will stop bacterial growth and
reduce autolytic activity, but the moisture content must be reduced to 15 per
cent to prevent mould growth. Salt retards bacterial action and aids the
removal of water by osmosis. When fish are salted prior to drying, a final
moisture content of between 35 per cent and 45 per cent in the flesh,
depending on the salt concentration, is often sufficient to inhibit bacteria.

I.   Types of fish

In tropical waters the catch is typically mixed and may include 300 or
more species all of which can be consumed fresh or in processed forms. Each
of these species will have different characteristics which will affect their
handling properties. The fish preserver working in the tropics may therefore
have to be prepared to preserve a range of species of different size and fat
content. Some of the species will be delicate and require careful handling
while others are more robust and less subject to damage.

The most important factors which affect the handling properties of a fish
are:

1.   size. Very small fish may be dried whole, whereas larger fish must
     always be cut open so as to increase the surface area available for salt
     penetration and/or moisture loss. Small fish may therefore be preserved

with the gut content intact, while this is almost always removed in larger species.

2. oil content. Fish oils oxidise readily and become rancid giving a bitter flavour to the product. Some communities show a preference for slightly rancid fish although rancidity is usually considered objectionable. Fish with a high oil content are difficult to convert into good salted and/or dried products since the oil acts as a barrier to salt penetration and moisture loss.

3. flesh texture. Fish with firm or moderately firm flesh are relatively easy to handle. They can be cut without falling apart and the dried product can be transported without breaking up. Fish which have a very soft flesh tend to tear when attempts are made to cut them, and the dried products are very fragile and tend to break up during transport.

I.1 Small pelagic species

Small pelagic species include fish less than 25 cm long. Such fish form characteristic schools or shoals, hence a catch containing these species will be comparatively homogeneous in composition. The group includes the herring-like and sardine-like fish which are slender and have relatively small scales and delicate flesh. Many species have a high oil content. These fish are sometimes dried whole without salting, but the products are then fragile and break easily. Rancidity is difficult to control, and unless the products can be marketed soon after drying, they will become progressively undesirable and unmarketable.

The small mackerels, such as the Indian chub mackerel, are also included in this group. These are sold fresh wherever possible, but may also be salted and dried. This group also includes anchovies and anchovy-like species of fresh waters. In Africa, near inland waters these are often sun dried without salting.

I.2 Large pelagic species

The most important fish in this group are the tunas which can attain weights in excess of 500 kg. Many other species reach a weight of 100 kg while some seldom reach 10 kg. The flesh is generally very firm and contains moderate amounts of oil.   In some species the flesh is very dark and many of these bleed heavily when cut.   The skin of most species is thin.

Most of the world tuna catch is canned but substantial amounts are sold fresh.   Fresh tuna is often highly priced and is not commonly used to make dried salted products.

The large mackerels and horse mackerels, or jacks, have moderately firm flesh and a medium oil content.   These are best sold fresh when possible, but good quality dried salted products can be made.

I.3 Small demersal species

Small demersal fish include bottom-living fish less than 25 cm long. They constitute a very diverse group including fish of very different shapes, but generally of deeper form than pelagic fish.   Most have quite large, hard scales and moderately firm flesh.   Some, such as the catfish, are scaleless and have soft flesh.   The oil content is variable and is generally less than 5 per cent.   There is less annual variation in oil content than is found in pelagic species.

The group includes many different types of sea-fish such as small mullets, snappers, sea-breams, croakers, jew fish and silver bellies, and small freshwater fish such as carps and breams.

Salted and dried products of good quality can be made from many small demersal fish but the products fetch generally low prices.   However these products are useful in that they provide lower-income groups with a source of animal protein food.

I.4 Large demersal species

Large demersal fish also constitute a diverse group. They include the sharks and rays, as well as bony fish such as mullets, snappers, groupers, jew fish, breams and threadfins.   Many of these bony fish are sold most profitably when fresh.   However they can also be processed into excellent dried salted products when demand for the fresh product is not sufficient. Good salted dried products can also be made from sharks and rays.

The freshwater fish in this group incude tilapia, carp and catfish. These species are sometimes split or cut into pieces and dried in the sun without salting.

II. Pre-processing stages

Fish spoil very quickly and small-scale fish-processing enterprises can easily lose produce and, hence, income through avoidable wastage. A great deal of spoilage may occur before the fish is processed. The bacterial and chemical changes which cause spoilage proceed rapidly at tropical temperatures. In general, the lower the fish temperature, the lower the amount of spoilage. It may also be reduced if fish are handled properly and good hygienic measures are adopted. A few measures for avoiding or minimising spoilage are briefly described below.

(i) Improvement of landing facilities and distribution

Very often whenever large catches are taken, landing facilities and the distribution system are inadequate to handle the surplus. In these circumstances a long period of time may elapse before the fish can be processed, with the consequence that a high percentage of the fish may be spoiled. The ideal solution to this problem would be to increase the amount of cold storage facilities. One simple method of reducing such a bottleneck is to develop the drying facilities as close to the landing areas as possible to reduce transport time and cost.

(ii) Maintaining the fish at low temperatures

If tropical fish are well chilled with sufficient ice, they may remain in an edible form for up to three weeks, depending on the species. In many locations ice may not be available to the small processor, in which case the fish can be kept cool by other means including:

- keeping the fish in the coolest spot available, such as in the shade;

- placing damp sacking over the fish. As the water evaporates from the cloth it helps to keep the temperature of the fish down. The sacking must be kept wet and the fish must be well ventilated;

- mixing the fish with wet grass or water weeds in an open-sided box so that the water can evaporate and cool the fish. With this method, the fish should be kept continuously wet.

(iii) Maintaining a hygienic environment

Fish which have been handled cleanly and carefully will be in better condition than fish which have been handled carelessly; they can, therefore, be worth more money.

Before processing starts, attention should be paid to the following points:

- keep the fish as clean as possible. Washing with clean water will remove any of the bacteria present on the fish skin, especially if it is muddy;

- keep the fish cool, chilled in ice or chilled water, if possible, at all stages before processing starts. Fish spoilage is a continuing process: once a particular stage of spoilage has been reached no amount of good practice or processing can reverse it;

- avoid damaging the fish by careless handling. If the skin is broken this will allow bacteria to enter the flesh more quickly and spoilage will be more rapid. This sort of damage can be caused by walking on fish and by the use of a shovel. If the guts can be removed and the gut cavity washed carefully this will reduce the number of bacteria; however, in some areas, the purchaser requires whole fish, and this practice may lower the value of the catch.

III. Processing techniques

III.1 Fish preparation

It is important that fish for salting and drying be prepared in a way which allows rapid salt penetration and water removal. Very small fish are sometimes processed without any preparatory cutting with only the guts removed whenever necessary. Fish larger than 15 cm are split open so that the surface area is increased and the flesh thickness is reduced. With fish more than about 25 cm long, additional cuts or scores should be made in the flesh. Depending on consumer preference, the head can be left on - or removed. It is desirable to scale fish for easy salt penetration and drying, however again the preference of the consumer must be considered.

Fish should never be prepared at ground level since it will pick up dirt even if placed on a board or mat. A table or bench at comfortable working height should be used. The table can be made out of wood, metal or concrete and should have a smooth surface which can be easily scrubbed clean. A separate wooden cutting board should be used on this surface to cut the fish. This will prevent either the knives or the surface being damaged. Good knives are essential for fish preparation.

Short knives should be used for small fish, long flexible knives for filleting and stout knives for splitting big fish. Knives must be kept sharp. Blunt knives tear the fish and slow down the work. If a grind stone is available, it should be used to shape or profile the cutting edge and to remove nicks. An oilstone or water-lubricated stone may then be used to sharpen the cutting edge. A steel should be used to remove burrs on the edge. Proper grindstones are expensive and steels are not easily obtained in some countries. In any case, a fish processor should always have a good sharpening stone available.

III.2 Gutting and splitting

Different techniques for opening and cleaning fish are used depending on the location and the species. The method used should be the one which gives a product which is recognisable and acceptable to the consumer. In all cases the guts, gills and hearts should be removed cleanly. Any dark coloured blood should be cleaned out using a small brush, and all black membranes should be removed from the inside of the fish.

III.3 Salting

Whether or not to salt before drying depends on location, availability of salt and, of course, consumer preference. Salting reduces the possibility of spoilage before, during and after drying and gives a stable product of higher moisture content than unsalted fish with a reasonable shelf life in humid conditions. The salt and extra moisture in a salted dried product gives the

cer more weight to sell than would be available from a comparable amount
nsalted dried fish.   Marine fish destined for drying are normally
d.   Salt is usually readily available at coastal locations and the
rvative effect it supplies may be essential when sun drying fish with hot
air.   At freshwater fisheries, salt may not be readily available or may
elatively expensive.   At these locations the preference may be for
ted dried fish.   If the ambient air at inland tropical locations is hot
ry, the fish can be dried to the lower moisture content and storage will
nt less of a problem.   Where unsalted dried fish are transported to more
areas care should be taken that the fish do not absorb moisture from the
nd become spoiled.

Generally only small fish should be dried unsalted, as larger fish will
before the drying process is completed.

There are three main salting methods: kench salting, pickle curing and
ing.   The first two methods yield fish with a relatively high salt
entration, while the third method (brining) is commonly used for products
a low salt concentration.   A method used in some fisheries, whereby fish
rubbed with salt and then hung to dry, is not recommended as it does not
uce an even cure.

Kench salting

In kench salting, the fish are mixed with dry crystalline salt and piled
the brine which forms as the salt takes water from the fish being allowed
drain away.   This method is especially popular for large lean fish
ies.   Kenching can be carried out in shallow concrete tanks fitted with a
n, or on raised platforms or racks of approximately 1 m^2 area and 8-10
ff the ground.   Starting at the centre of the rack, 2 or 3 rows of
ared fish are laid flesh side up over a bed of salt.   Salt is then
nkled or rubbed all over the fish, more being put on the thick parts of
fish than on the thin parts.   Wherever scores have been made, these
ld be filled with salt.   A pile of fish is built up by moving outwards
the centre, and sprinkling each layer of fish with salt before covering
the next layer.   To ensure good drainage, the centre of the pile should
bout 10 cm higher than the outside edges and the pile should not be higher
about 2 m.

Care should be taken in making the pile in order to ensure even salting of the fish and a good product quality. Brine should not be allowed to accumulate as this will produce an uneven cure and may discolour the fish. The edges of the kench pile should also be regularly sprinkled with salt to prevent contamination.

In the tropics, fish are usually left in the kench pile for 24 to 48 hours after which they are dried. However, the salt may not have completely penetrated the fish during this time, and penetration may continue during drying. In rainy weather, the fish may be left in the kench pile for longer periods. In this event, the pile should be broken down and a new pile made up, so that the top fish from the first pile are placed at the bottom of the new pile. In making the first kench pile, 30-35 parts by weight of salt should be used for each 100 parts of fish.

The advantage of kench salting is that the fluids are drained off leaving the flesh fairly dry. However, it also has a number of disadvantages: oily types of fish become rancid due to exposure to the air; insects and rodents have ready access to the fish; mould and bacterial attack can take place; and salting may not always be even.

Pickle curing

In pickle curing, a barrel or tank is used to hold the brine which forms as the salt mixes with the water contained in the fish. From 20 to 35 parts by weight of salt to 100 parts by weight of fish may be used depending on the cure required. Fatty fish, such as mackerel, are commonly pickle-cured.

In this salting method, a layer of dry salt is spread over the bottom of the tank upon which the first layer of fish is laid. There is, however, no need to stack fish higher in the centre as drainage is not required. The layers of salt and fish are stacked up, care being taken to ensure that no fish are overlapped without a salt layer between them, since this could cause the fish to stick together. As the pile is built up, the salt layers should become thicker. The top layer of fish must be placed skin side uppermost. A wooden cover should be placed on this top layer, so that weights can be used to keep the fish below the surface of the brine which forms.

Pickle curing is recommended in preference to kench salting as it produces a more even salt penetration and provides a better protection of the fish against insects and animals, since the fish are covered with brine.

3. Brine salting

In brining, or brine salting, the fish are immersed in a solution of salt and water. By varying the strength of the brine and the curing period, it is possible to control the salt concentration in the final product. The method is commonly used in developed countries when a smoked product is to be made and the salt concentration required in the final product must be lower than 3 per cent (e.g. as for hot-smoked mackerel). Brine salting may be used to advantage in developing countries as the process is more uniform and controllable than the dry salting techniques.

A fully saturated brine contains about 360 g of salt to each litre of water (3 lb 10 oz of salt per imperial gallon). A sack of salt should be hung in the brine to ensure that the latter remains at full strength. Full strength or saturated brine is called a 100° brine. A 10° brine - which is made up by mixing 1 part of 100° brine with 9 parts of water - is sometimes used to soak fish before salting.

4. Salt quality

The salt used for curing fish (fishery salt) is a mixture of a number of chemicals. A good fishery salt contains from 95 to 98 per cent common salt, known chemically as sodium chloride. Since fishery salt generally originates from the sea, it contains impurities such as chlorides and sulphates of calcium and magnesium, and sodium sulphate and carbonates. Other types of fishery salt include rock salt (i.e. mined salt) and sun salt or solar salt (i.e. salt obtained through water evaporation from coastal lagoons or ponds).

The type and quality of salt used affect the appearance, flavour and shelf life of cured fish. If pure sodium chloride is used for curing, the product is pale yellow in colour and soft. A small proportion of calcium and magnesium salts is desirable, as the latter yield a whiter, firmer cure which is preferred by most people. However, if the proportion of these chemicals is too high, the rate at which the sodium chloride impregnates the fish is slowed down. Furthermore, the salt becomes damp as the chemicals absorb moisture from the air making the product taste bitter.

The composition of sun or solar salt is determined by various factors outside the control of the processed-fish producer. Therefore, if salt from one source proves unsatisfactory, another source should be sought or the curer should consider making his/her own salt.

Solar salt often contains some sand and mud, as it is usually scraped up from the bottom of the ponds in which it is made. The cheapest grades contain a large proportion of dirt and these should not be bought for fish curing. Salt should be kept in clean bags or covered bins so that it does not become dirty.

Salt may also contain both moulds and bacteria. The bacteria cause the pink colour sometimes seen in salted fish. These bacteria also make the fish slimy and produce an unpleasant odour. If the salt is kept in storage under dry conditions for 6 to 12 months, the number of bacteria present will be much reduced. Alternatively, the salt can be baked to kill the bacteria. Both storage and baking will increase the processing costs. These may be avoided if some consumers of traditional products prefer the strong flavours produced in cured fish by mild attacks of pink bacteria.

All processing equipment and surfaces must be thoroughly washed with fresh water to help prevent pinking. Light growths can be brushed off from the fish surface and the product redried, but severe attack leads to the destruction of the fish.

Solar salt often contains very large pieces which should be ground up before use. An ideal salt for dry salting operations contains some very fine grains which will dissolve quickly and some larger ones which will dissolve more slowly and prevent the fish from sticking together. Very fine salt is preferred for making brines, because it dissolves quickly.

III.4 Drying

Fish are subject to the basic principles of drying described previously, i.e. water is removed from the fish by evaporation in two phases. During the first phase, when only water on the surface of the fish or very close to the surface evaporates, the rate of drying is mainly dependent on the rate of passage of hot air over the fish and on the ability of that air to absorb moisture. The drying rate can be raised by increasing the fish surface area by splitting, scoring and so on.

During the second phase, or falling rate period of drying, evaporation from the fish surface will be dependent on the rate of moisture transfer from inside the fish to the surface. Moisture transfer rates will be lower for oily fish since the oil acts as a barrier to moisture movement. In common with all commodities the second phase of fish drying will depend on piece size (i.e. distance from the surface), temperature and the amount of residual water.

It should be noted that if fish are dried too rapidly a hard impermeable outer crust will form which will prevent the passage of any more moisture. This phenomenon is known as case hardening. Externally, case hardened fish look well dried but the centre of the fish will still be moist and could spoil. A fish which has been damaged in this way will be hard on the outside, but may feel soft or spongy internally when pressed.

Fish are rich in proteins which are denatured when heated. This means that the fish might start to cook at conventional hot air drying temperatures. Optimum drying temperatures for temperate salt-water species can be as low as 27°C, whereas tropical species can generally withstand higher temperatures before cooking and can be dried using air at 45-50°C.

1. Conventional sun drying

Natural drying methods use the combined action of the sun and wind without the use of any equipment. Since it is important to dry the fish quickly before they spoil, all the fish surfaces should be exposed to the drying action of the wind. Ideally the drying site should be in a breezy location with a dry prevailing wind coming from an inland direction.

Traditionally, many fishermen in tropical countries spread their fish on the ground, on rocks, or on beaches to dry in the sun. Some fish processors use mats or reeds laid on the ground to prevent contamination of the fish by dirt, mud and sand. Drying fish in this manner has many disadvantages and, in recent years, the use of raised sloping drying racks has been introduced as a simple, but often effective, improvement. The product obtained from rack drying is cleaner since the fish do not come into contact with the ground; they are also less accessible to domestic animals and pests, such as mice, rats and crawling insects, which contaminate or consume them. Protection from rain is simply accomplished by covering the rack with a sheet of waterproof material (e.g. plastic); if fish on the ground are covered, they

are protected from falling rain but not from water on the ground itself.
Drying rates are higher, because air currents are stronger at a metre or so
above the ground and air can pass under the fish as well as over them. The
use of a sloping rack allows any exudate to drain away.

Where only a few large fish are to be dried, this may be done by hanging
the fish up. Split fish may be hung on hooks by tying them up with string,
or by tying the fish in pairs by the tail and hanging them across a pole or
line. Bombay duck is an example of a fish dried in India in this fashion.
The fish are hung in pairs, joined jaw to jaw, on horizontal ropes at an
optimum loading of 50-60 fish per metre.

Large quantities of fish should be dried on racks (figure 7). Suitable
materials for drying racks include chicken wire, old fishing nets, and thin
rods or poles, such as reeds or sections of bamboo. The surface of the racks
should be at a height of about 1 m from the ground and should slope if split
large fish are to be dried. A flat surface is preferred for drying small
intact fish. Where large quantities of very small fish are to be dried, a
netting rack may be impractical. Suitable drying surfaces may be made
instead, with raised floors of wood, concrete, bamboo strip or, where none of
these materials are available, well-compacted clay.

At the end of the drying day the fish are usually heaped together and
stored to prevent them becoming wet by dew or rain during the night. During
overnight storage the fish should be covered by wooden boards weighted to
apply pressure on the fish. This will flatten the fish, giving them a better
appearance, and will also speed up the process by which water moves from the
inside of the fish to the outside, so that they will dry more rapidly when set
out the following morning.

However, even when racks are used, sun drying has many limitations: long
periods of sunshine without rain are required; drying rates are low; and in
areas of high humidity, it is often difficult to dry the fish sufficiently.
The quality of sun-dried fish is likely to be low due to slow drying, insect
damage, and contamination from air-borne dust. Also it is difficult to obtain
a uniform product.

2. Solar drying

Thus, in the search for improved drying techniques, the use of solar

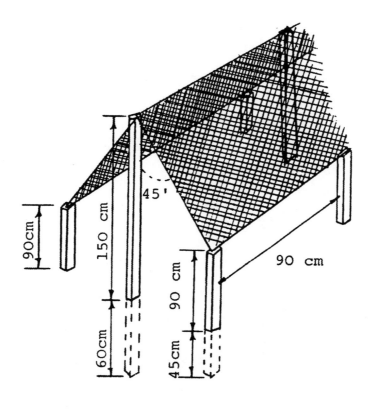

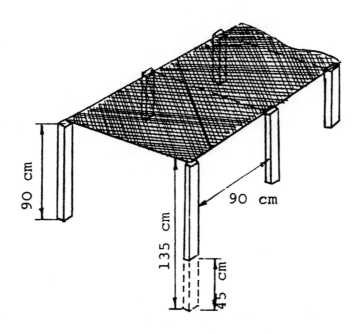

Figure 7

Fixed drying racks with flat and slanting drying surfaces

dryers has been investigated as an alternative to traditional sun drying. As previously discussed, solar dryers employ some means of collecting or concentrating solar radiation with the result that elevated temperatures and, in turn, lower relative humidities are achieved for drying. When using solar dryers, the drying rate can be increased, lower moisture contents can be attained, and product quality is higher. The dryers are less susceptible to variations in weather, although drying is obviously slower during inclement weather, and they do provide shelter from the rain. The high internal temperatures discourage the entry of pests into the dryer and can be lethal to any which do enter.

The solar tent dryer described earlier was first developed in Bangladesh for fish drying. This and several further modifications of it have since been tested in many parts of the tropics. With an ambient temperature of $27^{\circ}C$, temperatures of around $48^{\circ}C$ can be attained inside the dryer, which are ideal for tropical fish drying. The fish should be loaded onto racks built inside the solar tent in a similar fashion to sun drying.

In fine weather conditions fish can be dried within 3 days, compared with 5 days for sun drying. The quality of the solar dried fish is higher. During the initial constant rate period of drying (dependent largely on air movement), drying rates in the solar tent and on sun drying racks are broadly similar. However the second phase, or falling rate period of drying, occurs more rapidly within the tent. A suitable method for fish drying might be to use racks, in the first instant, and complete the process inside the solar tent. This would have the advantage of reducing the time spent by a batch of fish inside the dryer, allowing a greater throughput.

Solar drying reduces the effect of insect infestation on fish. In addition to causing losses in quality and quantity, insect pests are potential carriers of pathogenic bacteria and thus represent a serious health hazard. The temperatures found in solar dryers can kill any insects or larvae present on the fish, thereby presenting a means of disinfestation. A period of 20 hours at $45^{\circ}C$ is recommended for a complete disinfestation of drying fish.

Flies, the major carriers of disease, lay eggs on fish during the early stages of drying, but become less attracted to them as the flesh dries and hardens. The larvae tunnel into the flesh, causing putrefaction and extensive physical damage. The most important pests of the dried fish are beetles of the family <u>Dermestidae</u>. They invade the fish flesh from the earliest stages of drying, but unlike flies, continue to be attracted to, and breed in, the dried product.

Prevention of spoilage of fish during processing

A number of measures may be adopted in order to minimise the spoilage of fish during processing. These include:

- keeping all tools, fish boxes and cutting surfaces clean. Where drinking quality water is available, it should be used to wash the fish before and during processing, such as after gutting or splitting the fish;

- prevention of fish offal (guts, heads, gills, etc.) from coming into contact with cleaned fish. In addition, the fish working area should be cleaned regularly, at least once a day, by removing all offal and dirt which might contain bacteria or attract insect pests such as flies. All offal should be removed from the working site. It may be used as fertilizer, or buried. It should not be thrown into the water near the work site, as this practice fouls the water and may attract insects;

- ensuring that high standards of personal hygiene are maintained. Fish processors are handling food, and hands should always be washed before starting work and particularly after visiting the toilet. People who have infected wounds, stomach complaints, or any contagious disease should not be allowed to handle the fish;

- speedy processing. The longer the time required for processing, the greater the amount of spoilage which will occur before processing is completed;

- keeping the fish in boxes and off the ground. Work, such as cutting fish prior to salting or drying, must be carried out on tables, not on the ground where the fish will become dirty and pick up bacteria;

- protection of the fish from rain and using salt during drying in order to avoid the spoilage of fish through bacteria, mould or insect attack;

- protection of the fish against insect infestation during processing. Blowflies lay their eggs in the fish while they are still moist and the larvae eat the flesh. Beetles, such as the hide beetle, lay eggs in the fish as they are drying and the larvae eat the flesh even when it is quite dry. Damage can be reduced by ensuring that processing waste is properly disposed of so that there are no places for insects to breed. Use of better salting techniques may help since insect larvae are not attracted by heavily salted fish. Techniques which speed the drying process are useful in countering blowflies.

IV. Packaging

Dried fish are sometimes brittle and easily damaged if not handled correctly. In humid conditions, dried fish also absorb moisture and become susceptible to spoilage by moulds and bacteria. They may also be attacked by insects (especially beetles of the genus Dermestes), rats and mice, as well as domestic animals. Packaging methods such as hessian sacks, wooden boxes, and baskets are generally inadequate in protecting dried fish from these causes of damage.

To protect dried fish properly, one of the following measures should be adopted:

- packing of the dried fish in a sturdy container, such as a wooden or cardboard box, fitted with a lid in order to totally enclose the product. Open boxes, although protecting the fish from physical damage, are not effective against high humidity and insect attack. Properly sealed cartons, made from waxed or plastic-coated board, should be sufficiently moisture-proof and rigid enough to withstand rough handling. Although this type of packaging is more expensive than traditional packaging, the additional cost should be more than offset by the decrease in the spoilage rate;

- packing of fish in plastic or polythene bags, thus reducing insect attack and the effects of high humidity. Care should be taken not to leave bags containing dried fish in direct sunlight or in hot places, since the increased temperature causes "sweating" (i.e. the removal of water still present in the dried fish). This water condenses on the inside of the polythene bag, wets the dried fish and makes them susceptible to mould attack. A further disadvantage is that some dried fish have sharp, hard points and edges, which puncture and rip the plastic or polythene bags, thus allowing air, moisture, dust, or insects to spoil the fish.

A suitable arrangement might be to use a polythene liner within a close weave basket. Disinfestation of stored dried fish can be achieved by fumigation. Since the chemicals used for fumigation are also toxic to humans, extreme care is necessary when fumigating any products. Experienced and trained personnel should carry out the process. Phostoxin and methyl bromide are effective fumigants. Fumigation should be carried out in an enclosed fish store or under gas-proof sheets in order to ensure a complete disinfestation of stored fish. 24 g of methyl bromide per cubic metre has been found to disinfest dried fish successfully when applied over a 24 hour period. However, phostoxin is considered more suitable for use in fish stores at a dose of 0.2 to 0.5 g phosphine per 50 kg for 2 or 3 days. Dried fish can also be disinfested by heat treatment, which can be supplied by a solar dryer as mentioned earlier.

V. Alternative processes

The reader should be aware that simple processing of fish is not restricted to drying. Other products which can be made include a variety of smoked and cured goods and also fermented fish products. These techniques may also be appropriate for use in the small-scale fisheries sector. Information on these processes can be found in some of the references listed.

VI. Extension work activities for fish drying

VI.1. Location

Fresh fish do not travel well in hot climates. Therefore for the purpose of demonstrating fish drying to extension workers, it is preferable to find a site near fish landing areas. Any national institute or agency involved in fisheries should be able to advise on a suitable location. It may be possible to work side by side or in cooperation with an on-going project, which should reduce the effort required to set up a demonstration unit and assist in the procurement of the raw materials. Any agency already working in fisheries development may also be able to recommend suitable extension workers and advise on their educational background.

VI.2 Current practices

Within the chosen location, the present fish drying practices should be established:

- what species (or groups) are dried?
- are the fish salted before drying?
- are the fish filleted, gutted, etc.?
- have any difficulties in production been recognised?
- who dries the fish?
- what are the uses of the product?
- does the market require a specific quality of product?
- what is the size of the market?
- are there any shortfalls/gluts in production?
- how much fish does a processor dry, and how does this relate to overall production?
- what route does a processor's output take to reach the market? (If it is bulked with the dried fish made by other people it may be difficult to introduce changes.)

The technologist should listen carefully to how the processor and the purchaser of the dried fish describe the product. It is unlikely that either will be totally objective, but it may be possible to infer whether an improved processs would give a better product acceptable to all. The technologist will be aware of any deficiencies in product quality because of his formal training, but these may not be immediately obvious to the layman.

VI.3 Demonstration units

In order to demonstrate the techniques to the fishermen it is essential that the extension workers become fully conversant with the methods the technologist is recommending. This is equally true whether you are attempting to introduce fish drying as a new technique or trying to improve on existing techniques. A demonstration unit should be constructed where all the processes associated with good fish drying can be carried out. Samples of dried fish produced in this fashion will illustrate the efficiency of the process to extension workers who will then be able to transfer the information to fishermen at a later date.

For the purpose of demonstration, a process line of about 10-20 kg capacity is sufficient. A small unit will help reduce costs and by keeping the time spent at each unit operation (gutting, salting, etc.) to a minimum, ensure that the demonstration does not become tedious.

Prepare a list of all training materials required:
- work surfaces, knives, fish boxes, clean water, etc. for cleaning the fish;

- salt and any containers necessary to salt the fish;

- wood (prepared or otherwise), carpentry tools, netting, wire mesh, nails, etc. needed to construct solar dryers and drying racks;

- plastic sheet to cover the solar dryer. The best way to obtain this will vary from location to location. Plastics manufacturers might not want to sell you a small amount, in which case you may either be given the amount you require or be sold a larger amount. If the latter is the case, you might be able to recover costs by acting as a central plastics store for any subsequent solar drying extension work. Polyethylene is also useful for forming into plastic bags for packaging;

- paint and preservatives to treat the wood;

- instruments to monitor drying. The number of these you have will depend on local availability and the extent of your budget. Spring or two-pan balances, and thermometers (wet and dry bulb) should be available in most principal cities;

- packaging materials: bags, boxes, baskets, and cartons;

- a lockable tool box. Useful items such as balances, knives, hammers, etc. have a tendency to be damaged or on occasion borrowed and not returned. In order to conserve your stock it is necessary to keep them under lock and key, for use only with your authority.

Security is also important when deciding on the location of the demonstration unit. The preparation techniques such as cleaning and gutting, and possibly even salting, may be best carried out indoors. The construction of drying racks and solar dryers will be done in the open. A secure site where the drying work will be seen and also protected from theft or damage will be needed. The drying site should of course be adjacent to the fish preparation area. It is essential to establish who will be responsible for maintaining the site and any drying material built on it. Where practical, the flat roof of any suitable building would be an ideal location.

Construct drying racks and solar dryers of an appropriate scale, making careful notes of the construction techniques and raw materials used. If any method seems impractical or unnecessarily complicated, improve upon the design.

Using the equipment available, carry out fish drying exercises until you are familiar with methods of gutting, salting, drying, etc.. Prepare dried fish at a level of quality you consider appropriate for the market, and obtain some opinions on your product from the trade. Having established that your development is acceptable, you are now in a position to make recommendations to extension workers.

VI.4 Involvement of extension workers

If possible, arrange for suitable extension workers to visit your demonstration units for training programmes lasting 1-2 weeks. Groups of 10-20 workers could be trained at a time depending on the level of interest shown and the manpower available.

Discuss the current fish drying practices, bringing their attention to spoilage and infestation. Then, you should describe the processes you have developed and explain the reasons for any modifications to traditional practices. The extension workers should carry out a programme of practical sessions in order to familiarise themselves with sun and solar drying processes and the construction methods used for the dryers. Encourage them to describe the techniques in their terms and in those understandable by artisans or fishermen/women. Carry out a simple costing exercise to indicate the cost of the drying units, and discuss whether this amount is feasible in relation to the price which can be obtained for the dried fish. By the end of the training programme the extension worker should have assimilated all of the necessary technical information and should be able to convey this to the end-user of the technology. He/she should also be capable of undertaking all the operations associated with the technology, from fish gutting to dryer construction, and be confident of the reasons for carrying out these tasks. He/she should be aware of all the advantages and disadvantages of each technique. The adoption of the technology now depends on the identification of a target group of artisans or fishermen/women by the extension services, where the approved methods can be demonstrated by the trained extension force. The technologist should be retained in an advisory capacity during this phase, to provide technical back-up as necessary.

CHAPTER 5

VEGETABLE DRYING

Vegetables are a desirable component of any diet, because they provide the complete range of nutrients required. Their production is often seasonal depending on annual rains and suitable growing temperatures. There are likely to be periods when surpluses are available which the farmer may not be able to use or sell profitably. At other times of the year there are likely to be shortages which can result in the variety and nutritional content of the diet being restricted. This is particularly evident in low income groups. The introduction of simple processing techniques to allow vegetables to become available all year, will go some way to alleviating this problem and will also help to raise the incomes of rural people.

Drying is a particularly suitable method of preserving vegetables. When deciding on which vegetables to dry the reader should make the distinction between vegetables which may be traditionally dried or consumed fresh in the locality - in this group we may include okra, chillies and various herbs - and vegetables which may be eaten exclusively in the fresh form. The introduction of dried versions of the latter may meet with consumer resistance which should be taken into consideration in any extension proposal.

Certain vegetables such as lettuce, melons, cucumbers, radishes and asparagus are not suitable for drying.

Vegetables are not subject to attack by the same range of pathogenic organisms which thrive on flesh foods such as fish. Poor drying techniques are more likely to lead to spoiled produce rather than dried foods which are dangerous to the health. However the reader should be aware that mould growth on incorrectly dried vegetables (and most other commodities) can contain toxins which are poisonous.

I. Pre-processing techniques

The best dried vegetables will be produced from freshly harvested produce and not from the second class remains of a batch the farmer was unsuccessful in marketing fresh. It is essential to plan at the time of harvest how much will be sold fresh and how much will be dried. The vegetables to be dried should be kept cool, shaded and, in the case of leafy vegetables, sprayed with clean water to minimise wilting. Correct handling practices will minimise bruising and damage to the crop. Ideally the crop should be transported to the drying site in manageably sized boxes and baskets rather than moved in bulk in a large cart and shovelled off.

It is important that all post-harvest operations should be carried out as soon as possible after harvesting to minimise the extent of spoilage.

I.1 Hygiene

The need for good hygiene during all processing operations cannot be over-emphasised. Frequent hand washing, regular cleaning of equipment, and rapid disposal of waste materials go a long way in preventing contamination of the dried product. Adequate provision of preparation equipment, for example, sharp (stainless steel) knives; availability of potable (or chlorinated) water; plenty of room for working; and people who have been trained in processing procedures and the need for hygienic conditions; together with correct location of the processing site are other factors which contribute to hygienic working conditions.

I.2 Cleaning

Following harvesting and transportation to the drying site, the first pre-drying operation is usually cleaning, although this is not an essential operation for all commodities. Cleaning serves to remove dirt, leaves, twigs, stones, insects, insecticidal residues, and other contaminants. The removal of these contaminants reduces spoilage rates. The easiest way to wash the crop is manually in a purpose-built tank made from concrete or plastic, or in clay pots. It is essential to change the water regularly especially when washing heavily soiled vegetables such as root crops.

I.3 Grading and sorting

Sorting is very important in the production of high quality foodstuffs. However it should be noted that the operation can be made much easier (and losses reduced) if correct harvesting procedures are followed. Sorting may be performed at the time the commodity is received at the drying site, but is sometimes done immediately after cleaning when the physical characteristics of the commodity are better exposed. Factors that may be considered in sorting are size and shape, colour, texture, density, chemical compositon, blemishes, and insect infestation.

Sorting serves to determine the value of the commodity for processing and assists in making the operation more profitable.

Quality aspects such as maturity, colour, flavour, and the absence of defects govern the acceptability of the finished product. Therefore sorting is second in importance to correct harvesting for producing high quality dried products. During sorting, all the substandard produce should be rejected. Sorting is usually followed by grading, where the vegetables are subdivided into groups based on the desired quality attributes. This may be done on the basis of size. For small quantities this process can be done manually.

Grading into lots according to shape or size has the following advantages:

1.   any subsequent pre-processing operation such as blanching or sulphuring can be more easily controlled;

2.   drying will be more uniform;

3.   the market may pay more for an evenly sized product. In some cases a higher price may be obtained for a particularly desirable size at this stage;

Size grading is only relevant to vegetables which will be dried whole. Where the vegetables will be cut or sliced, size grading is carried out after this.

I.4 Peeling

Many vegetables require peeling prior to drying (see plate 5.1). Since a thick skin presents a physical barrier to moisture removal its removal aids drying. However, care must be taken not to remove too thick a layer in case

Preparation of vegetable

Plate 5.1
Peeling yams. Good
sharp knives and a
clean work suface
are essential

Plate 5.2
The peeled and sliced vegetables should be stored under water

able nutrients are lost, e.g. peeling a thick layer from potatoes or
er removes most of the vitamins and minerals.   Losses can also be high if
ing is not properly controlled.   Freshness is important; sometimes
velled or corky skins develop on old vegetables which require special
ing treatment.

In most cases manual peeling using good stainless steel knives or potato
ers is the cheapest and simplest method.

Coring, pitting and trimming

The object of this stage is to remove any undesirable part of the
table.   Any blemishes should be cut out.   In the case of root vegetables
green tops and long roots should be removed.   With green vegetables any
y stem or core should be cut out.

Cutting and slicing

Unless the vegetable is going to be dried whole, such as okra, it is
ssary to reduce its size to enable drying to take place in a sensible
.   Depending on the shape required by the market, vegetables may be cut
cubes, slices, strips, rings, shreds, or wedges.   Again cutting with a
p knife is the cheapest and simplest method.   Care should be taken to
in pieces with as high a degree of uniformity as possible.   Grading
ld be carried out.

All of the manual cutting operations described can be time consuming.   To
ent the prepared material from browning or deteriorating, it can be stored
orarily in a 2-4 per cent brine solution until the next batch processing
ation (see plate 5.2).   In the case of green leafy vegetables, the
ing gives an improved texture which makes the vegetables easier to dry.

Blanching

Blanching involves subjecting vegetables to boiling or near-boiling water
eratures for very short periods.   The principal function of blanching is
inactivate plant enzymes, but the operation also partially cooks the
ues and renders the cell membranes more permeable to moisture transfer.
rapid and complete drying is obtained and the texture is improved when

the blanched, dried product is rehydrated. During blanching, the micro-organism count in foodstuffs is substantially reduced and sometimes the food becomes practically sterile. However, during exposure to the non-sterile conditions that can exist in subsequent processing operations, the product can easily become recontaminated.

All of the operations described, from harvesting to slicing, result in damage to the plant tissue. The response to this by the plant is the release of enzymes which can produce undesirable quality changes. For example, when potatoes are cut and exposed to the air, enzyme activity will result in the formation of undesirable brown pigments. Other than discolouration, the undesirable effects of enzyme activity are the development of off-flavours, the loss of vitamins, and the breakdown of tissue.

Enzyme systems in foodstuffs are extremely complex and vary with different commodities. Most of the enzyme systems are progressively inactivated at temperatures above $70^{o}C$ and completely deactivated at around $90^{o}C$. Sometimes exposure of one or two minutes in the blanching vessel is sufficient, but longer times are usually used to ensure complete inactivation. Peroxidase is one of the most heat-resistant enzymes. The test for the destruction of this enzyme can be used to indicate the adequacy of blanching in the factory or laboratory.

During the blanching process, the time of exposure to the heating medium required for a given commodity depends upon several factors:

(i) piece size. To obtain effective enzyme inactivation all parts of the material should reach a temperature of at least $90^{o}C$. Longer blanching times are required for larger pieces to allow penetration of heat to the centres. However care should be taken in the case of large pieces that their outer surface does not become cooked;

(ii) temperature. Suitable uniform temperatures must be maintained throughout the blancher. In mountainous regions, where the barometric pressure is lower, immersion times must be increased to compensate for the lower boiling temperature of water;

(ii) depth of load. Heat must penetrate into the centre of the bed of material so that all pieces will reach the desired temperature;

(iv) blanching medium: Blanching in water usually requires less time than blanching in steam at the same temperature, because of the rapid application of heat to each piece in the liquid medium, as against the relatively slow penetration of steam into the food.

Blanching can either be carried out in water or steam. For water blanching, the commodity is immersed in a container of boiling or near boiling water for the necessary time. Care must be taken to avoid over-blanching which leads to loss of texture and difficulties in subsequent drying. On a small scale the easiest method is to use a muslin bag or wire basket for immersion which not only ensures the same immersion time, but keeps all the pieces together in a clean environment.

The product to be blanched can be placed on a muslin cloth, the corners of which are then tied to give a simple bag (see plate 5.3). The bag can then be attached to the end of a stick and dipped into a boiling pot. Care should be taken not to put too large a batch in the boiling water or it will cool the water excessively. The water temperature in the blancher should not drop below about 90°C and, if possible, should be rapidly brought back to the boil after immersing the vegetables (see plate 5.4). With pieces of 6 mm thickness adequate blanching can be obtained after about 4-6 minutes using this method. While the bag is in the pot it should be stirred around to ensure even heating. After blanching the bag should be rapidly cooled using cold water to prevent over-blanching or cooking.

Blanching by immersion has the advantage that comparatively large amounts of material can be processed at one time. However a high level of soluble solids can leach into the water from the vegetable, which may reduce the nutritional content or render the dried food less attractive.

Steam blanching produces a more attractive dried product and is often preferred to water blanching because there is a smaller loss of nutrients by leaching and, in some vegetables, the dried product has an enhanced storage life. It basically consists of subjecting the prepared commodity to an atmosphere of live steam. If the commodity is on trays, therefore, there need be no direct handling of the commodity between blanching and drying, thus aiding hygiene. On a small scale, this can be carried out by placing the tray in the upper section of a tank or pot, the lower section of which contains vigorously boiling water, and covering with a lid. The tray should not come into direct contact with the boiling water.

Plate 5.3
Blanching can be done
in a cooking pot over
a charcoal stove

Plate 5.4
Ensure that the water
temperature stays high

As previously mentioned, steam blanching times are usually a few minutes longer than water blanching times. Vegetable pieces of 6 mm thickness would be adequately blanched after 6-8 minutes. These times are given only as guidelines. Trial and error will determine the most suitable blanching time under local circumstances.

I.8 Colour retention

Blanching can be used as a tool to aid colour retention, especially in green vegetables. The green colour is due to the presence of chlorophyll which breaks down when green tissues are heated in the presence of acid. During blanching, since the contents of plant tissues are acid, the pH of the blanching water falls, bleaching the leaves. When green vegetables such as cabbage are blanched, it is necessary to use a sodium carbonate solution to maintain the pH of the liquor at 7.3 to 7.8 to prevent discolouration. A practical method of doing this is to use a 1 per cent sodium bicarbonate solution for blanching, using the immersion technique.

Blanching is not carried out with all vegetables. For example onions and garlic are not blanched, since this would result in a loss in pungency as the flavour compounds are very volatile and would be lost through blanching.

I.9 Sulphuring and sulphiting

With some dried products the use of chemical preservatives will improve the colour and increase the shelf life. The most common preservative used is sulphur dioxide. There are two methods of providing sulphur dioxide (SO_2) to commodities, sulphuring and sulphiting. Sulphuring is more commonly used for fruits, and sulphiting for vegetables.

Sulphiting involves introducing SO_2 into the commodity by the use of sulphite salts such as sodium sulphite or sodium metabisulphite, either by adding them to the blanching water or, when steam blanching is employed, by either spraying a sulphite solution on to the commodity, or by soaking it in a cold solution following blanching. Blanching in a sulphite solution is particularly useful since it combines two operations into one. The concentration of sulphite salts, and dipping, spraying or blanching times again depend on the commodity.

Concentrations used for spraying applications should be between 0.2-0.5 per cent. Sulphiting by the use of sprays or dips in not generally thought to be practical in the rural sector.

Where the chemicals can be obtained, immersion blanching in a sodium metabisulphite solution might be practical. It is important to control sulphiting accurately to obtain the correct levels of SO_2 in the food. Excessive amounts of SO_2 give the food an unpleasant smell and may cause the food to be unacceptable.

The strength of a sodium metabisulphite solution is expressed in parts per million (ppm) otherwise known as mg per kg. By way of conversion, 10,000 ppm of SO_2 means there is an overall concentration of 1 per cent. For example, 1.5 g of sodium metabisulphite dissolved in 1 litre of water will give 1,000 ppm (0.1 per cent) of SO_2. An easy way to prepare a stock solution is to dissolve 12 g (2 1/2 level teaspoons) in 1 litre of water to give an SO_2 strength of 8,000 ppm (0.8 per cent). By adding more water to this, different strengths can be achieved. Some examples of the use of a sulphiting solution are given in the techniques for drying vegetables below.

Sulphuring involves burning elemental sulphur in a sulphur chamber to produce SO_2 which permeates into food tissues. A sulphur chamber consists of an enclosure, with adjustable vents, housing perforated trays stacked one above the other. The amount of sulphur used and the time of exposure depend upon the commodity, its moisture content, other pre-treatments and the permitted levels in the final product. The food is placed on the trays inside the sulphur box and sufficient sulphur is placed in a container near the trays. For most vegetables 10-12 g of sulphur (2-2 1/2 level teaspoons) per kg of food is adequate. The sulphur is ignited and allowed to burn in the muffled environment for 1-3 hours.

Sulphuring has the advantage over sulphiting that it uses rock sulphur, which may be more readily available than sodium metabisulphite. The fumes of burning sulphur are unpleasant and can be dangerous to the processor. Sulphuring should always be carried out outdoors in a well ventilated location.

I.10 Quality advantages

Other than the discolourations caused by enzymes described for blanching, browning reactions caused by chemical reactions can occur during drying.

These adverse quality changes are referred to as non-enzymic browning. They are particularly noticeable in light coloured foods such as potato and cauliflower. The rate of most browning reactions is greatly dependent upon temperature, the combined effect of time and temperature, and the moisture content of the drying material. The chemical reactions involved occur in water and proceed slowly in the dilute solutions found in very moist foods. As the foods are concentrated during the drying process, the reactions proceed more rapidly.

The browning reactions continue after drying, and dried vegetables will continue to darken during storage unless they have been treated with SO_2. The presence of SO_2 in the dried vegetables will also inhibit microbiological spoilage and will help to deter insects both during drying and later in storage.

It should be emphasised that SO_2 is toxic at high concentrations and its use should be carefully controlled. Some countries have mandatory limits on how much SO_2 can be included. The limits for some dried fruit and vegetables in the United Kingdom are shown in table 2. If the dried product is intended for export, the possibility of national regulations should be borne in mind. In some countries, such as the Federal Republic of Germany, the use of SO_2 is not permitted.

Table 2
Maximum permitted levels of additives for
some dried foods in the United Kingdom

Food	Preservative	Level (ppm)
Coconut, desiccated	Sulphur dioxide	50
Fruit:		
Figs	Sulphur dioxide	2,000
	Sorbic acid	500
Prunes	Sulphur dioxide	2,000
	Sorbic acid	1,000
Others	Sulphur dioxide	2,000
Vegetables:		
Cabbage	Sulphur dioxide	2,000
Potato	Sulphur dioxide	550
Others	Sulphur dioxide	2,000

I.11 Quality changes during drying

In order to select the correct drying conditions it is useful to be aware of some of the factors which affect vegetables during drying.

The product may be adversely affected by light. Sunlight triggers a whole range of undesirable reactions. It can cause oxidation of fats which causes rancidity. Light can initiate undesirable colour changes; green vegetables may become bleached, and pale coloured vegetables can darken. Sunlight can also cause a reduction in the nutritional value; vitamins can be destroyed. One method of overcoming these problems is to dry the vegetables in the shade wherever possible.

Excessively high temperatures during drying can result in high levels of shrinkage in the dried foods. This may render them irregular in shape and size, and unattractive to the consumer. High temperatures will also increase the tendency towards browning. Dried vegetables which are badly shrunken are more difficult to rehydrate.

The rehydration of a dried food product is another important factor, particularly in terms of consumer acceptance. It must not be thought that rehydration is a complete reversal of the dehydration process. Some of the changes produced by drying are irreversible. The outer layers of the dried food, crushed and malformed during drying, are unable to return completely to their original size and shape. Soluble constituents in the dried foods leach out into the rehydration water contributing to a loss in nutrients and also a loss in flavour and aroma.

The swelling power of starch and the elasticity of cell walls, both important rehydration factors, are reduced by heat treatment. With some products, such as dried fish, producing a rehydrated product similar in appearance to the original form is not so important, as the dried product is utilised in soups or stews and is added directly to the cooking pot. This may also be true for dried vegetables.

II. Choice of drying technique

II.1 Sun drying

The sun drying of vegetables on the ground should be avoided to prevent contamination by dust and insects. Where sun drying on the ground is the practice, attempts should be made to introduce simple improvements such as the use of mats. However, for the same reasons as described for fish drying, it is preferable to raise the drying vegetables off the ground by the use of trays on racks. Vegetables can be dried on horizontal racks. The trays should be perforated to permit the maximum flow of air around the drying vegetables. The trays should be loaded with no more than 6 kg of vegetables per square metre. The vegetables should be spread in an even layer and should be stirred or moved at least every hour during the first drying period. This will speed up drying and improve the quality of the finished product. Sun drying of vegetables on racks is illustrated in plates 5.5 and 5.6.

Shade drying should be carried out if it is necessary to prevent discolouration or to conserve nutrients. In this case the drying rack should be placed in a shaded position. Shade drying is more dependant on air movement through or over the drying vegetables. Therefore the drying rack should be positioned to take maximum advantage of any winds. In dry air conditions with ample circulation, shade drying can be accomplished almost as quickly as sun drying. In conditions of high sunshine and low humidity sun drying can be finished in one daylight period. Where a longer period is necessary the same precautions concerning protection from evening rains and morning dews, described for fish drying, apply.

II.2 Solar drying

Most vegetables can be satisfactorily dried using a solar cabinet dryer. During the initial stages of drying it is essential to ensure that there is no condensation of water inside the dryer. Condensation is caused by insufficient air flow and the operator should ensure that the air intake and outlet vents are sufficiently wide open to prevent this happening. Increased air flow inside the cabinet will reduce the temperature, but the reader will recall that the initial drying phase is more dependant on air flow than high temperature. The alternative, with no air flow, is a hot steamy cabinet which acts more as a blancher than a dryer. In the second phase of drying the air flow through the cabinet can be reduced allowing the temperature to rise. Cabinet temperatures in the range of $60-80^{\circ}C$ are adequate to dry most vegetables.

Sun drying

Plate 5.5

Sun drying potato slices on a raised tray

Plate 5.6

Sun drying tomato slices and onion rings. The trays are raised upon
racks but these two foods should not be placed side by side as shown here

As described previously, the solar cabinet dryer can be used in the shaded mode by placing a black absorbant material directly below the cover. This absorber will transmit most of the heat received to the air inside the dryer. When drying more than one vegetable at a time it is essential to ensure that the correct combination of different varieties is placed side by side on the rack or in the dryer. For example, strongly flavoured vegetables such as onions and chillies should not be placed next to more bland foods such as tomatoes, otherwise, the tomatoes will pick up the strong flavour. Tent drying of tomatoes and solar cabinet drying of carrots are illustrated in plates 5.7 and 5.8.

III. Specific vegetable drying techniques

The following methods have been collected from various sources in different countries and give an idea of what processing techniques might be suitable. The reader may find it necessary to modify them to suit local practices.

III.1   General method - green legumes

After shelling, the outer skin of the pea or bean is gently punctured to assist the internal moisture to escape. This can be done by tapping a single layer of the legume under an open meshed tray. The legumes are then scarified by brushing a stiff-bristled steel wire brush lightly across the top of the mesh or by gently tapping the brush on the peas, allowing the bristles to penetrate slightly. Blanch by immersion in boiling water containing 1 per cent sodium bicarbonate (2 teaspoons of baking soda per litre) for 2 minutes. Spread evenly in a single layer and dry at 55-60°C until the legumes are brittle and crisp. Shade drying may help preserve colour.

III.2   Specific methods

Green soyabeans. Steam blanch the pods for 10-15 minutes until the beans are tender but firm. Shell and dry until brittle.

Green beans. (i) Remove strings from string varieties and split the pods lengthwise to hasten drying. Steam blanch for 15-20 minutes and dry until brittle.

Green beans. (ii) Cut the vegetables into 2 cm pieces discarding the end of the pod. Blanch in 1 per cent sodium bicarbonate solution for 4-6

Solar drying

Plate 5.7

Tent drying of tomatoes

Plate 5.8

Cabinet-dried sliced carrots

minutes. Cool and dip the beans for 1 minute in a sodium metabisulphite solution containing 8000 ppm SO_2. Dry until brittle, equivalent to a moisture content of about 6 per cent.

Okra. (i) Okra may be dried whole or in halves or strips after blanching (6 minutes in boiling 1 per cent sodium bicarbonate solution). After blanching remove the slimy exudate. Dry in a single layer at $60-65^{\circ}C$, preferably in the shade.

Okra. (ii) Cut young tender pods into 1 cm pieces or split lengthwise. Steam blanch for 5-8 minutes. Dry in a bed not more than 1 cm deep until very brittle.

Peas. Shell young tender pods and then immediately steam blanch for 5-8 minutes. Dry, stirring frequently, until the peas are hard and wrinkled.

III.3   General method - green leafy vegetables

The vegetables should be washed, stems removed as necessary and the leaves cut into strips about 5 mm wide. Blanch by dipping in boiling 1 per cent sodium bicarbonate solution for 3 minutes. Sulphite (optional) by dipping in an 8000 ppm SO_2 solution for 1 minute. Spread the vegetables thinly and dry quickly, preferably in the shade, at $55-60^{\circ}C$. Drying may be achieved in as little as 4 hours.

III.4   Specific methods

Broccoli. Trim, cut, wash, etc. Steam blanch for 8-10 minutes and dry until crisp.

Cabbage. (i) Remove outer leaves, quarter, core, and cut into shreds. Steam blanch for 8-10 minutes. Dry until tough and leathery.

Cabbage. (ii) Prepare as for method (i), then blanch in a boiling 1 per cent solution of sodium bicarbonate for 3 minutes, followed by a 1 minute dip in an 8,000 ppm SO_2 solution. Dry to a moisture content of about 5 per cent.

Cauliflower. Separate into florets. Dip in salt solution (2 tablespoons of salt per litre of water). Steam until tender. Dry until the florets are hard to crisp and tannish yellow in colour.

Celery. Strip off leaves, cut into 1 cm pieces. Steam blanch until tender. Stir occasionally during drying. Dry until brittle.

Parsley, Jews-Mallow/molochea and herbs. No pre-processing necessary. After washing, spread loosely and shade dry, or alternatively, hang bunches of the whole plant in a dry warm shady place. Drying should be complete in a few hours. When dry and brittle crush the leaves and remove the stems.

Spinach. Select young tender leaves, wash and cut large leaves into several pieces. Steam blanch for about 4 minutes until wilted. Dry until brittle.

III.5 Roots and tubers

Potatoes. (i) Wash, peel, trim, and cut into thin slices. To prevent browning, place in a 1 per cent solution. Blanch in boiling water for 5 minutes. Sulphite (optional) for 1 minute in an 8,000 ppm SO_2 solution. Spread slices thinly and evenly, and dry at 60-70°C until crisp and brittle, equal to a moisture content of 6 per cent.

Potatoes. (ii) Wash, etc., and cut into long strips of 5 mm cross section or slices about 3 mm thick. Rinse in cold water and steam blanch for 4-6 minutes. Dry until brittle.

Sweet potatoes and yam. As for potatoes.

Turnips and swedes. (i) As for potatoes.

Turnips and swedes. (ii) Wash, trim, etc. as before. Quarter, peel and cut into 3 mm slices or strips. Steam blanch for 15 minutes and dry until leathery.

Carrots. (i) After washing and scraping, cut into 9 mm slices and blanch for 5 minutes in boiling water. Sulphite (optional) by dipping in 8,000 ppm SO_2 solution for 1 minute. Drying conditions as for potatoes. Shade drying will help to preserve colour and pro-vitamin A content.

Carrots. (ii) Crisp tender carrots free from woodiness should be washed trimmed and cut into strips or slices about 3 mm thick. Steam blanch for 8-10 minutes and dry until tough and leathery, equal to a moisture content of 6 per cent.

Parsnips. As for carrots.

Beetroot. Small tender beets of good colour and free from woodiness are washed and trimmed. The beets are steamed for 30-45 minutes to cook through, then cooled, peeled and cut into slices or strips about 3 mm thick. The beet pieces are dried until tough, leathery or brittle.

III.6   Other vegetables

Onions. After removal of tops, roots and outer leaves, the onions should be washed and sliced into thin (3 mm) rings. They should not be blanched since this destroys flavour. They do not require sulphiting. The rings should be dried until crisp at 55-70°C and packed immediately in air- and moisture-proof containers. Onion rings are highly hygroscopic. The safe storage moisture content is about 5 per cent.

Leeks. As for onion.

Garlic. Separate the cloves and remove the outer skins. Finely chop the cloves into pieces smaller than 5 mm. Similar to onion, blanching or sulphiting is not necessary. Dry the pieces until the garlic is brittle, equal to a moisture content of 5 per cent. Separate off the dry skin by winnowing, and pack the garlic in air- and moisture-proof containers.

Peppers, sweet (capsicum). The washed, opened and cored peppers are sliced into thin strips. No blanching is necessary. Sulphite (optional) by dipping for 1 minute in 2,000 ppm SO_2 solution. Shade dry at 55-65°C until crisp, equal to a moisture content of 7 per cent.

Tomato. (i) Steam or dip in boiling water to loosen skins. Chill in cold water and peel. Cut into sections not over 20 mm wide. Sulphur (optional) in box for 10-20 minutes. Dry until leathery.

Tomato. (ii) After washing and trimming, slice the fruit and dip for 3 minutes in a solution containing 600 ppm SO_2 and 10 per cent salt. Dry the slices until they are leathery, equal to a moisture content of about 6 per cent.

Chillies. (i) Wash and dry whole without blanching or sulphiting. Shade dry at 60-65°C.

Chillies. (ii) Wash, trim, and cut into 1 cm strips or rings. Remove the seeds. Steam until tender, then dry the rings 2 layers deep until they are pliable.

Mushrooms. Peel larger mushrooms and dry whole or sliced, depending on size. If stems are tender, slice for drying, if tough, discard. No pre-processing is necessary. Spread not more than 1 cm deep, and dry until leathery to brittle.

IV. Packaging of dried vegetables

IV.1 Reasons for packaging

All dried foodstuffs are normally packaged in some way for storage and marketing. Whether the package is a large one, perhaps for distribution to an industrial or trade customer, or a small package for sale to a household consumer, three basic functions of the package can be recognised:

(i) it should contain the foodstuff, enabling the chosen quantity to be handled as one unit without loss, throughout the hazards of transport and storage;

(ii) it should protect the foodstuff and preserve its required attributes through a planned shelf life;

(iii) it should communicate information about the foodstuff such as its nature, origin, method of use, quantity, destination, and name of producer.

With dried vegetables the need to protect the foodstuff is of primary consideration when selecting the method of packaging. The prevention of typical moisture contents in the range of 2-8 per cent and require packaging which gives good protection against moisture uptake. This is particularly so in humid tropical climates.

The tendency to pick up water, as water vapour, is determined by the equilibrium relative humidity of the storage atmosphere, not merely by the moisture content of the food. With some foods, such as potato crisps, moisture content relates directly to a critical quality attribute (crispness), and in conditions where the food gains moisture during storage, the shelf life is simply the time taken to reach a critical moisture content. With many other foods the moisture content influences the rate of deterioration in quality through chemical and biochemical reactions or through microbiological spoilage.

High temperatures may have adverse effects on foods, but packaging cannot provide direct protection except in the very short term. Oxygen is a significant factor in deterioration of many foods, especially those containing fats or oils, and this can be taken into account in the packaging system. Light may have effects similar to those of oxidation, or may promote oxidation, and the package can be selected to exclude light from the food.

This is particularly important where processing steps, such as the use of sodium bicarbonate or shade drying, have been taken to preserve the colour of green vegetables. Any quality advantages which have been gained by these steps will be lost if light is not excluded from the packaged goods.

The package should also prevent tainting of the food by foreign odours. Good storage practise will also help reduce the likelihood of odour contamination. Strongly flavoured dried vegetables such as onions should not be kept next to blander products which would pick up the onion flavour.

A good package will also help to retain the SO_2 content of the dried vegetable. This will help retain the good appearance of the product and enhance the shelf life of the package.

Biological hazards to dried foods need attention throughout production and distribution, although they are usually a secondary consideration in relation to selection of a package. Bacterial spoilage will not be a problem in a dried food, and the action of a package is simply to exclude contamination, thereby preventing any increase in the bacterial load. However, if the package is unable to prevent the absorption of moisture by the dried vegetables, mould growth will occur which will spoil the food.

Infestation by mites or insects is a potential problem with most dried foods; packaging has a part to play in control, but will only be effective if combined with other measures. Protection against mammals (especially rodents) and birds cannot be provided economically through packaging, although good packaging practice will assist control through minimising spillage. These pests must be excluded from food stores, by appropriate store design and management procedures.

IV.2 Types of packaging

Traditional packaging materials used for long established products such as dried fruit and dried fish include baskets, bales, jute sacks, wooden boxes and cardboard boxes. These can only be used for packaging dried products which do not require water vapour or oxygen barriers under the prevailing climatic conditions. They are appropriate for commodities which are transported in large packages to a central marketing point and then sold loose. These packages can often be used several times and are usually cheap; but the need for good hygiene must be emphasised. Inclusion of a low density polyethylene film lining in a traditional package can add substantial protection against uptake of water vapour.

A wide range of protective packaging materials has been used for dried foodstuffs including metal cans, glass jars, and rigid moulded plastic containers, as well as the more commonly found flexible packs. Rigid packages may be assumed to have negligible permeation rates for water vapour and gases, if they are properly sealed, but they are relatively expensive and of limited relevance to the needs of the small-scale processor in developing countries.

A wide range of flexible packaging materials is available including paper, aluminium foil, plastic films, and also laminates incorporating two or more of these. Again the relevance of many of these is limited with the possible exception of polyethylene (low or high density) which may be manufactured locally, and is cheaply available. Low density polyethylene, while not the optimum dried-food packaging material because of its high permeability to oxygen, has two advantages. It is a moderately good moisture barrier and it can be easily sealed in a candle flame or by using aluminium foil and a soldering iron.

Flexible materials may be used as the sole component of a small package or as a barrier component in a package. A sturdy outer package is also required in each instance; this could comprise a basket, a fibreboard or wooden box, a paper wrap, or a textile or paper sack. This outer package provides much of the protection against transport hazards, and the choice therefore influences the strength required in the inner pack. In practice, therefore, the complete package system should be planned as a whole. It should be noted that in many circumstances the choice of packaging materials is limited, and it may be a case of utilising whatever material is available. The packaging of dried onions and tomatoes using simple plastic bags, a cheap and effective way, is illustrated in plates 5.9 and 5.10.

IV.3 Alternative processes

The reader should bear in mind that there are other methods for preserving vegetables. If the main purpose of preservation is to extend the period of availability, this can also be done by curing, in the case of onions, or simply by good storage practice, in the case of root vegetables. Other appropriate technologies which could be considered include salting and pickling. These are traditional methods with widespread applications in the rural sectors of many countries.

V. Extension work activities for vegetable drying

This will be discussed in the following chapter on fruit drying since the suggested techniques are similar for both commodity groups.

Packaging

Plate 5.9
Dried onion rings
The food should be
packed as soon as
possible to prevent
any moisture re-
absorption

Plate 5.10
Dried tomato slices.
Polyethylene bags provide
a reasonable barrier
against moisture and can
be sealed in a flame

CHAPTER 6

FRUIT DRYING

Fresh and dried fruit contribute a range of micronutrients to the diet and are particularly valued as sources of ascorbic acid (vitamin C) and beta carotene (pro-vitamin A). Fresh fruit consumption is often limited by seasonal availability and in some cases by high cost. Strangely, in some cultures fruit is considered to be almost a luxury item and consumed in small amounts whereas in others it is a dietary staple. Any steps which result in increasing the availability of fruit in deficient areas can only be considered as meritorious. In this context fruit drying can provide a means of increasing availability.

Two forms of dried fruit exist - semi-moist fruits and dried fruits. Semi-moist fruits, such as dried grapes, contain naturally high levels of sugar. This means that dried grapes can be preserved at a higher moisture content than other dried foods. Typically, semi-moist products can have a moisture content of about 25 per cent. This gives these products the advantage that they can be eaten directly in their preserved state without any need to rehydrate them. There is considerable world trade in semi-moist fruit, particularly dried grapes from Australia, the United States and Cyprus. The distinction should also be made between traditionally dried fruits, such as dates and apricots, and novel dried fruits such as papaya or in some cases mango. It will almost certainly be more difficult to sell dried mango than dried apricot to the uninitiated. Most fruit varieties are suitable for drying with the exception of citrus fruit, although there is a limited production of sun-dried limes in the Middle East.

I. Pre-processing techniques

Some fruits have a tendency to ripen rapidly, after which they start to deteriorate in eating quality. As with vegetables, it is essential that only the best produce be used for drying, and since the correct harvest period for fruits might be quite short, there is a greater time constraint placed on the processing methods used. There is considerable variation in the ripening characteristics of different fruit. Mangos and bananas ripen rapidly and can have short harvest periods whereas citrus ripen slowly.

All fruit to be dried should be hand picked and not shaken from the tree. This will prevent bruising. The fruit should be picked when just ripe and ready to eat. Over-ripe fruit will bruise easily and may be soft and difficult to slice. The copious amount of juice in some over-ripe fruit makes it sticky and difficult to handle. During subsequent processing, over-ripe fruit can absorb an excess of SO_2 during sulphuring which renders the dried product dark and unattractive. Conversely, under-ripe fruit may not absorb sufficient SO_2 and will give a hard, bitter, unattractive dried product. There is therefore little sense in preparing and drying fruits which have been harvested at the wrong time.

All of the information on the pre-processing of vegetables is pertinent to fruit with the exception that fruit are normally not blanched. Browning can occur when the flesh of fruits such as apples or bananas is cut. To prevent this the fruit should not be cut up until the last possible minute, and the cut pieces should then be stored under water until time for the next processing stage. It is customary to sulphur fruit, using a sulphur box rather than to use a sodium metabisulphite dip. Sulphuring is essential to preserve the light colour of some dried fruits such as apples and apricots. Sulphuring also helps to preserve the ascorbic acid and the beta carotene content during drying. As previously mentioned, it has the additional advantages of controlling microbial and insect activity and of protecting delicate flavours. Generally, about 3.5 to 4 kg of sulphur per tonne of fresh fruit is sufficient. This is about half the level recommended for vegetables.

II. Pre-processing methods specific to fruit

II.1 Checking

Some fruit with a waxy skin is dipped in a hot dilute caustic soda solution prior to sulphuring. This process is called checking. Typically, washed grapes (or plums) are dipped in a 0.3 per cent boiling sodium hydroxide solution for 3 or 4 seconds. This produces cracks in the fruit skin which speed up the subsequent dehydration process. The grapes are then washed in cold water to prevent any more chemical damage and to avoid cooking the product. Another cold water dip is used to eliminate all traces of sodium hydroxide from the skin. Where checking is not a feasible proposition, due to the lack of sodium hydroxide or a reluctance to use it, an alternative is steam blanching for 10-15 minutes. The fruit is then sulphured.

II.2 Sugaring

There are various methods of applying sugar (sucrose) to fruit. The simplest is to apply a dusting of fine sugar just before drying. This may help retard browning and give the dried product a sweet coating.

Fruit pieces can also be dipped in a concentrated syrup. This results in the movement of water out of the fruit by the process of osmosis. At the same time, a smaller amount of sugar will penetrate the fruit tissue. This phenomenon of osmosis, the diffusion of a solvent through a semi-permeable membrane from a dilute to a concentrated solution, proceeds until an equilibrium concentration is reached. The solute (sugar) is able to diffuse through the membrane in the reverse direction only very slowly, so that the net result is transfer of water to the concentrated solution.

By immersing fruit in a concentrated sugar solution, water equivalent to over 50 per cent of the initial fruit weight can be removed, thereby greatly reducing the load on the dryer in the subsequent solar drying stage. It should be noted that the product obtained is different to that obtained by drying alone. The process is analogous to salting fish. In the same way that saltfish do not require to have as much moisture removed as unsalted dried fish (because of the preservative effect of salt), fruit which has been osmotically treated will be shelf stable at high moisture contents and have textures similar to semi-moist fruits such as raisins. This is due to the preservative effect of the additional sugar absorbed from the syrup. Some of the advantages of including this osmotic step in the drying process are:

(i)   during osmosis the material is not subjected to a high temperature over an extended time, so heat damage to colour and flavour are minimised;

(ii) a high concentration of sugar surrounding the material prevents discolouration by enzymic or oxidative browning. A good colour can be obtained in the dried product without chemical treatments, such as sulphiting;

(iii) as water is removed by osmosis, some of the fruit acid is removed along with it. This lower acid content, combined with the small amount of sugar added to the fruit during osmosis, produces a blander and sweeter product than ordinary dried fruit.

Some of the disadvantages are:

(i) a thin film of sugar is left on the surface of the fruit after drying, which may be undesirable. However, this can be removed by a quick rinse in water after osmosis;

(ii) the process produces a dilute syrup as a by-product. This can be brought back to strength by concentrating or by adding more sugar, and recycled. However, there is a limit to how often it can be used before becoming unacceptably contaminated. Suggested uses for the syrup have included fruit nectars;

(iii) the process may be unnecessarily complicated by including this step;

(iv) sugar may be expensive.

The concentration of solution and time of soaking are dependant on the material and the desired level of water removal. The technique has been tested with banana, mango and papaya. Suggested soaking times are up to 18 hours in a 67 per cent sucrose syrup, stirring occasionally. This will remove about 40 per cent of the original moisture. This soak is followed by a one hour soak in a 60 per cent sugar solution containing one per cent SO_2 (as sodium metabisulphite), and finally a rinse in cold water to remove subsequent stickiness. The fruit can then be sulphured and solar dried. Shade drying will help to conserve the colour.

A third method of introducing sugar is to immerse the fruit pieces in a boiling syrup for a few minutes. This will result in a change in the texture of the fruit, and a hard gel will form on the surface. A partially candied fruit product can be made in this fashion which can then be sulphured and dried. Care should be taken not to immerse the fruit in the syrup for too long or else extensive browning will occur. If over-ripe fruit is used, it will tend to disintegrate in the syrup.

The advantages and disadvantages of this process are the same as for the osmotic process described above.

Unlike most of the pre-processing operations described, sugaring is not essential to give a good dried product. The technique has been included as an example of an additional operation which _may_ be useful. The reader must decide its relevance to his/her particular situation.

III. Drying methods

Most of the countries which dry fruit commercially have hot sunny climates which favour sun drying methods. Indeed exposure to the sun is essential to obtain the correct colour development in grape drying. Even in developed countries with access to sophisticated alternative techniques, sun drying of fruit is still the principle method of drying since it gives the desired product quality.

The drying methods described for vegetables are entirely appropriate for fruit drying. The simplest method which can be recommended is sun drying on racks. It is essential when drying fruit to keep it off the ground to speed the drying process and also to reduce theft or damage by predators. Drying fruits are particularly attractive to insects and animals and it is desirable to keep losses to a minimum because of the comparatively high value of the crop.

Generally speaking, fruit takes longer to dry than vegetables. This is due to the sugar content which can make the drying fruit sticky and retard the rate of moisture loss. Semi-moist fruits, such as dates, have a high invert sugar concentration. Invert sugars and fructose in particular, are hygroscopic. In a hot dry climate, sufficient moisture will be lost to give the desired semi-moist product with good keeping qualities. However, in humid climates the fruit may not lose sufficient water quickly enough to prevent spoilage. It is possible in some cases that drying fruit exposed to a sudden increase in humidity will reabsorb moisture from the air. Where fruit with a high sugar content cannot be dried sufficiently fast, moulds and yeasts are likely to grow which will ferment the sugars present.

In areas where sun drying is not feasible, due perhaps to high humidities, solar drying using a cabinet dryer is a possible alternative. Operating conditions are similar to those described for vegetables. When drying fruit which is valued for its vitamin content, such as mango, it is worth considering shaded cabinet drying. On the other hand where exposure to the sun is necessary, as in grapes, direct cabinet drying should be carried out.

III.1 Some specific fruit drying methods

Suitable drying temperatures for fruit are 60-80°C.

Banana. The fruit should be ripe, and sweet, but not soft or brown. Cut into thin slices 5-7 mm thick, and sulphur. Alternatively sulphite by dipping in a 2000 ppm SO_2 solution for 1 minute. Dry the fruit in a single layer at 60-75°C until it is hard and brittle, equal to a moisture content of about 12 per cent. Avoid overheating the banana to prevent darkening.

Breadfruit. Peel, core and cut into chips or thin slices. Dry as for banana.

Apples. Peel, core and cut into slices or rings. Sulphur for 60 minutes and dry until the fruit is leathery and has no moist area in the centre.

Pears. Peel, cut in half lengthways, core, and form slices about 3-5 mm thick. Sulphur for 60 minutes and dry until the fruit texture is springy.

Peaches. Peel carefully and avoid bruising. Dry as for pears until pliable, but leathery.

Apricots. Cut the fruit into half and pit. Apricots will dry more rapidly if quartered or sliced but check that smaller pieces are acceptable in the market. Sulphur for 60 minutes. Dry until pliable, but leathery.

Plums. Cut in half and pit. Check and sulphur for 60 minutes. A handful of plums properly dried will fall apart after squeezing.

Berries. Wash and check. Sulphur for 60 minutes. Dry until the berries are hard and there is no visible moisture when crushed. Strawberries are not suitable for drying.

Figs. If the figs are small or have been partly dried on the tree they may be dried whole without checking or blanching. Otherwise cut in half, check and dry until pliable and leathery, but still slightly sticky.

Dates. Dates may be partially or wholly dried on the palm depending on the climate. Where they are partly dried they can then be sun or solar dried whole without any pre-processing. Direct sunlight is essential. Alternatively the dates can be pitted, halved and sulphured before drying. The semi-moist date-halves can be pressed together to form a paste.

Grapes. Wash, check and sulphur for 60 minutes. Dry until pliable and leathery. Seedless varieties are preferable for drying.

Mango. Peel and cut off the two fleshy cheeks. Cut into thin slices. Treat with sugar (optional) and sulphur for 60 minutes. Shade dry. Well dried mango will be golden brown and pliable. Different varieties of mango have different drying quantities. If excessive browning occurs, dry at lower temperatures.

Papaya. Dry as for mango.

The maximum permissible moisture content for the safe storage of some dried fruit is shown below.

Table 3

Fruit	Maximum permissible moisture content (per cent)
Raisins	25.0
Sultanas	20.0
Prunes	21.5
Apricots	20.0
Peaches	18.0
Dates	24.0

IV. Packaging

The packaging requirements of many dried fruits are similar to those of dried vegetables. However, the semi-moist products in particular have special packaging needs to prevent the reabsorption of water. Since dried fruits are a valuable commodity it may be possible to spend more on the packaging material. A moisture-proof, sealed plastic bag inside a cardboard box would be a suitable package. It is likely that the producer of dried fruit will be in an area where there are other growers. It might be possible to arrange for communal packaging facilities. This would minimise costs and could result in a standard package (with the producers own identification if so desired) which would be easily recognised by the consumer.

V. Alternative processes

Other methods of processing fruit include preserve manufacture to give jams, jellies and chutneys. Fruits rich in sugar can be fermented to produce wines. Starchy fruit such as bananas can be sliced and fried to give chips.

VI. Extension work activities for fruit and vegetable drying

VI.1. Location

Fruit and vegetables are highly seasonal and may well be harvested in different parts of the country at different times of the year. It may be difficult to find one single location where all or even a representative cross section of the crops will be grown and harvested at a convenient time for drying. Consultation with horticultural agencies should assist in finding the most suitable spot to set up drying demonstration units. As with extension work for fish drying, it may be convenient to work in co-operation with other post-harvest development projects, sharing overheads and providing complementary skills. Horticultural units may also be able to provide information on the types of fruit and vegetables which are grown and on the varieties suitable for processing.

VI.2 Current practices

Determine which fruits and vegetables are traditionally dried and what techniques are used. Establish whether any improvements to the process are desirable and who would benefit from the improvements. Consider whether any crops which are usually consumed fresh could be dried and what the dried product could be used for. For example, tomatoes could be dried and sold as a direct replacement for tomato paste at the village level. Chilli peppers could be dried and sold either whole or ground up. It might be possible to sell dried chilli profitably in large towns where the present supply is poor or non-existant.

Determine the major season(s) for each fruit and vegetable crop which you are going to dry. Work out a processing schedule which will enable you to dry different commodities at different times of the year. In this way, through time you will build up a collection of various dried foods to show to extension workers. When extension workers come for training, it is unlikely that you will be able to demonstrate drying techniques for more than a handful of commodities, but the dried samples with an explanation on how they were made will provide a broader picture of the merits of the process.

VI.3 Demonstration units

Draw up a list of all the unit operations you are going to cover and the equipment you will need. For demonstration purposes, batches of up to 10 kg of raw material will be sufficient to teach the basic skills. The pilot plant can therefore be designed around this amount. The equipment you will need at each stage will include the following:

(i) preparation - knives, work surfaces, chopping boards, clean water, salt and containers for washing, peeling, etc.;

(ii) pre-processing - charcoal stove or other heat source, and pots for blanching and/or sulphiting; a wooden box for sulphuring; rock sulphur or sodium metabisulphite, sodium hydroxide, sodium bicarbonate, sugar;

(iii) drying - materials and tools for building drying racks and/or solar dryers; plastic material for covering the dryer and also for packaging; other packaging materials as appropriate;

(iv) monitoring - a set of scales, and wet and dry bulb thermometers are desirable.

As noted in the fish drying extension work section, it is important to guard all tools to prevent them being misplaced or damaged. They should be stored under lock and key when not in use and only issued on your authority.

The preparation stages can be housed indoors if a suitable building is available. The pre-processing and drying units should be constructed at a secure outdoor location where full advantage of the sunshine and any prevailing wind can be taken advantage of. The flat roof of a building would be a useful site.

Construct or collect the full range of dryers and processing aids you are going to test, keeping a record of any procurement difficulties experienced. Determine the cost of the pilot plant. A useful exercise at a later date might be to ask trained extension workers to cost comparative pilot plants using alternative materials.

Become familiar with each of the processing steps using a variety of fruit and vegetables. Retain samples of the dried products with careful records of the methods used to obtain them. Obtain market reaction to your products. Establish which processes and products are preferred and can be recommended to extension workers.

VI.4 Involvement of extension workers

Arrange for a group of extension workers involved in fruit and vegetable post-harvest technologies to attend a one to two week training programme at your demonstration site. During the training programme you should explain the reasons for drying crops and the quality advantages which can be gained by using the recommended processes. The trainees should be made aware of the advantages of:

- cleaning, sorting and grading;
- blanching - but discuss also the extra labour and cost incurred by using fuel;
- sulphuring/sulphiting - but consider the availability of sulphur-containing materials and the extra cost and effort required;
- shade drying;

- solar drying.

Compare the quality of sun dried with solar dried products and discuss which material the market would prefer.

The extension workers should have practical experience preparing dried foods and become familiar with the construction techniques used to produce the pilot plant.

Display your previously prepared dried foods and present illustrations on the methods used to produce it. Ask the trainees to suggest what other fruit and vegetables might be suitable for drying and get them to devise suitable processes. The trainees should present the information in a method understandable by the farmer. The trainees could also devise methods or recipes for using the dried foods.

Before the end of the training programme the extension workers should be aware of processing constraints, the most suitable methods to use and, bearing in mind the construction costs of any recommended method, what the quality and/or price advantages will be in the market.

Dissemination of solar drying technology in the Sudan: a case study

Using the techniques described above, a training programme - part of an ILO project on appropriate technological advancement in the rural sector of the least developed Arab countries - was carried out in cooperation with the food technology staff of the Food Research Institute, Khartoum. The target audience was extension workers active in the rural sectors of the province of Khartoum Province. The objective of the course was to increase awareness of sun and solar drying techniques and thereby provide the extension workers with more information to disseminate to farmers and artisans. The training programme was divided into two parts. In the first phase, solar dryers were constructed and the principles governing their use were demonstrated. This was followed by a practical session when a range of commodities was dried using different techniques and the quality of the dried goods evaluated.

Three types of solar dryers were built: the cabinet, the tent and the modified paddy dryer. For ease of construction, prepared timber and other conventional building materials were used. The extension workers soon became familiar with the construction principles. They were, however, less willing to suggest alternative building materials. This may have been because of unfamiliarity with rural materials or a subconscious belief that the expensive, prepared materials were best because of their professional appearance. The danger exists that this belief could subsequently be passed on to the farmer who could ill afford the luxury of expensive raw materials. Since most of the prepared timber, woodworking tools, nails, etc. are imported, any general recommendation to increase their consumption will have widespread implications. It is essential therefore that the extension workers should be made to question the appropriateness of any construction technique.

A supply of plastics to cover the dryers was a problem. The local plastic manufacturer was sympathetic and willing to sell us a relatively small amount. However, the factory at that time was not working due to lack of (imported) raw material. Fortunately, this situation resolved itself shortly afterwards.

After the practical drying sessions, the extension workers were asked about their opinions on the various products and processes. It was agreed that the solar cabinet dryer gave a good quality dried product and was relatively simple to construct. It was questioned whether the improved quality (in this case the exclusion of substantial amounts of dust) would be recognised by the farmer. In the rural areas concerned, dust was ubiquitous and people learnt to live with it. Because of this familiarity, a high dust content would not necessarily be considered a negative quality aspect. In any case the dried food would be exposed to dust in the market place. Packaging to exclude dust was not realistic since people liked to see and touch what they were buying.

Blanching and sulphuring/sulphiting were recognised as steps which improved the quality, but the extension workers did not consider them as realistic recommendations. The extra cost and effort required would not be reflected by a higher price in the market for the dried foods. The overall impression received from the extension workers was that until the consumer was prepared to pay a high price for improved quality (or alternatively not pay the asking price for inferior goods), there was little incentive to introduce improved drying techniques in their particular sphere of activity.

Dissemination of sun drying technology in Somalia : a case study

The ILO has been involved in income generating activities in refugee camps in Somalia. One project concerned the development of agriculture in Jalalazsi camp where refugees were encouraged to grow crops including tomatoes, onions, melons and chilli pepper. Many vegetables are in short supply in Somalia and it was considered that the farmers would be able to sell their surpluses in the surrounding towns and villages or in major population centres such as the capital, Mogadiscio. However, chilli pepper production was much greater than the demand for the fresh material. Therefore, it was decided that the project farm manager should introduce simple sun drying techniques. In the first instance chilli peppers were dried by placing them on mats on the ground. Sun drying took between 4 and 14 days depending on the weather. At a later stage of the project, drying on racks was introduced. The dried chillies were then ground into a powder using small hand-powered mills. Several problems were noted at this stage. If the chillis were incorrectly dried (due to humid conditions or insufficient time), they were soft and difficult to grind. The grinding work was unpopular since it could be arduous and the powdered pepper was an irritant. No grading of the peppers was carried out before grinding which gave an unevenly coloured powder. However, this observation was somewhat academic since local demand for the dried pepper far exceeded supply.

CHAPTER 7

GRAIN DRYING

I. Sun drying

Unlike most other foodstuffs, cereals dry out almost completely during the normal ripening process on the plant. Most areas of the world receive sufficient sunshine at the time of harvest, and dependence upon the sun for drying grain is the cheapest and most common practice. Cereals are routinely dried in bulk in this fashion. The question whether or not to dry grain very seldom arises since in most cases there is no alternative process. A notable exception is maize, which when eaten fresh as corn on the cob is usually considered a vegetable.

I.1 Millet and sorghum

Drying may be completed in the field before harvesting or else may be continued after the harvest by placing the crop in the sun. Sun drying can be completed before or after threshing. In the case of fine grains such as millet and sorghum, the seed heads can be spread out on mats or on a concrete floor. The grain should be threshed before putting it in store since closely packed grain is less subject to insect attack. Sorghum and millet grow in semi-arid areas and are therefore seldom exposed to the damp harvest conditions which prevent adequate drying and result in spoilage.

I.2 Rice

By contrast, rice,the staple food in the monsoon areas of Asia, is often harvested under humid conditions. Drying is necessary to reduce the moisture content to safe storage levels. Sun drying can be done before or after threshing. Unthreshed paddy is sun dried by leaving small bundles in the field for several days after cutting. Newly threshed paddy is sun dried by spreading it on mats or on a concrete floor. In some countries, areas of metalled roads are used. Periodic stirring is necessary to obtain uniform drying. Losses and contamination can be considerable. Variable weather conditions can result in poor quality, as rewetting and overdrying can crack the rice grains and produce a poor yield of full kernels after milling.

I.3 Maize

Cob maize, either husked or unhusked, is often dried in narrow cribs (figure 8) with open-wall construction to allow natural ventilation through the cobs. Under suitable climatic conditions, and provided that the air flow is not restricted, the cobs dry to a safe moisture content without the development of surface mould or insect infestation. These cribs are also commonly used for the temporary storage of dried cobs.

The sides of the crib can be made of wire mesh, or of any conventional material, such as loose-woven wattle, that does not obstruct the air flow more than the maize cobs themselves. Natural aeration will dry the cobs slowly but safely whether or not the cob sheaths are removed. Where the climate is generally dry after harvest, the crib may be at least 2 m wide. In humid areas, the width should be reduced to 1 m at most, and possibly to 60 cm.

For protection against rain, the roof should be as water-tight as possible. On the other hand, intermittent wetting of the sides of the cob mass will not seriously impede drying unless the wetting is prolonged and excessive. Extensive roof overhang at the sides is therefore usually unnecessary and has been shown to reduce the drying rate.

The posts and post-holes supporting the crib can, if necessary, be soaked with a persistent insecticide to stop termite damage. Rat damage is prevented by building the base of the crib at least 1 m above the ground. Rat-guards made of sheet metal cones, may also be mounted on the posts in order to prevent rats from climbing up the crib. These guards should stand out at least 25 cm from the post.

If birds represent a serious problem, additional screening should be used to keep them out. The mesh size of the screen should not exceed 2 cm.

Figure 8

Crib for storing maize grain delivered on the cob

Unhusked maize cobs are often tied by the tassels into small bundles and hung from trees or exterior house rafters to dry. Alternatively, simple racks, made from horizontal bars supported by inclined bamboo poles, may be used (figure 9). The cobs are hung from the horizontal bars by their tassels. In the event of a sudden shower, a similar but slightly taller rack covered with polythene sheeting or large leaves can be placed over the first rack (figure 10).

Maize grain shelled directly after harvest may be spread on the ground and sun dried to a safe moisture content. Drying under these conditions can be accelerated by placing the grain on a black surface, such as a black plastic sheeting. Two posts are knocked into the ground on either side of the plastic sheet and a rope or bamboo pole is suspended between them. The grain is spread over the sheet with the two ends of the latter left uncovered. In the event of a rain shower, the two ends can be lifted and pegged to the rope, thus forming a simple tent over the grain (figure 11). This type of drying system requires frequent mixing of the grain in order to prevent over-heating.

Many of the problems associated with grain drying can be attributed to the vast bulk which is handled. For example, most of the sun drying operations for both threshed grain and grain on the cob are conducted at ground level. In most cases it would be preferable to raise the cereals off the ground to improve air circulation and isolate pests, but where large quantities are involved it may not be practical to do so. Unthreshed grain can be dried in cribs in the fashion described for maize, but this method is not satisfactory for large volumes of threshed grains. This is because threshed grain forms a very compact structure which restricts the passage of air. Hence a large surface area is needed to turn and mix the grain. The small particle size of most food grains means that the crib material would have to be tightly woven to prevent leakage and this too would inhibit air flow.

II. Solar drying methods

Grains have a pronounced constant rate period of drying and therefore benefit from comparatively high air flow rates. Cereals can be damaged if dried at 60-70°C while rice should be dried at a lower temperature in the range of 45-55°C. Simple natural convection dryers such as the cabinet dryer

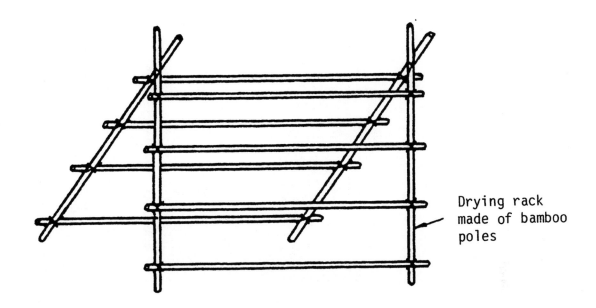

Drying rack
made of bamboo
poles

Figure 9

Simple rack for the sun drying of
unhusked maize cobs

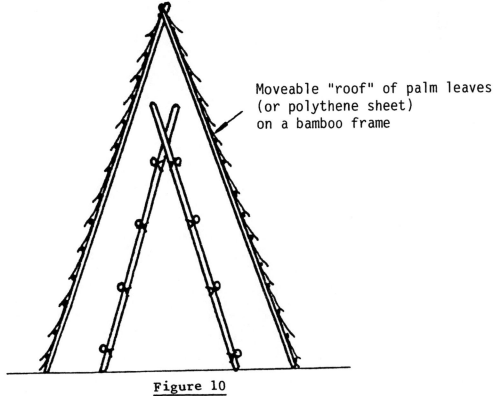

Moveable "roof" of palm leaves
(or polythene sheet)
on a bamboo frame

Figure 10

Drying rack inside a moveable "roof"

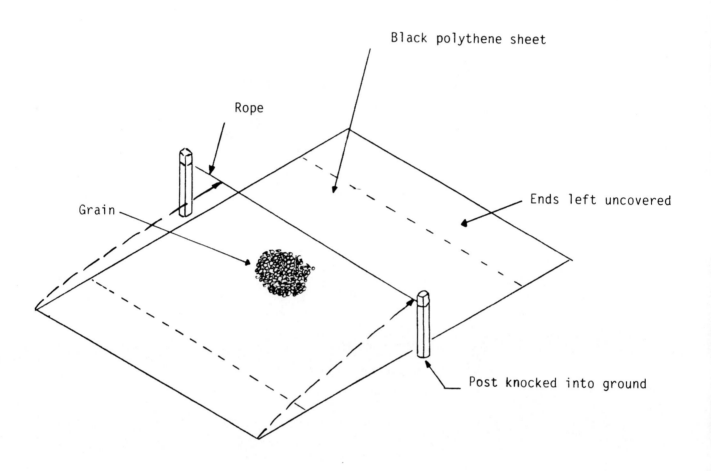

Black polythene sheet

Rope

Grain

Ends left uncovered

Post knocked into ground

Figure 11

Improved sun drying of shelled maize grain

(O'Kelly, 1979)

Table 4

Advised maximum moisture contents for

safe storage up to about 27°C

(Higher ambient temperatures require lower maximum moisture content)

Cereal	Percentage of moisture content (wet basis)
Wheat	13.5
Bulgur wheat	13.5
Wheat flour	12.0
Maize - yellow	13.0
- white	13.5
Maize meal	11.5
Milled rice	12.0
Sorghum	16.0

Note: The precise safe moisture content value differs to some extent between varieties for all cereals. The variation in grain sorghums is particularly great. The values given are the average for each particular cereal, and where there is any doubt about particular varieties, the equilibrium relative humidity of the grain, which must not exceed 70 per cent for safe storage, should be used as the deciding factor.

are not ideal for drying grain but should be used if available due to, for example, a gap in vegetable drying schedules. The maximum air flow possible through the cabinet (whilst maintaining the operating temperature in the desired range) should be obtained. Threshed grain should be spread in a layer 1.5 - 3.5 cm deep on drying trays of appropriate mesh size so as to give a loading of about 10-12 kg/m^2. For bulky materials, such as unthreshed sorghum, the layer can be up to 7 cm deep.

The natural convection dryer with separate collector and drying chamber (described under Group two dryers in Chapter 2) was designed as a possible solution to the problems of harvesting rice in the wet season in Thailand. The design consists of a solar air heater, a box for the rice bed, and a chimney giving a tall column of warm air to increase the convection effect.

Because of the increased air circulation, this design is said to be capable of satisfactorily drying paddy layers 10-15 cm deep within 3-4 days. It is recommended that the area of the solar collector should be about three times the area of the rice bed. The top of the bed should be shaded to prevent overheating of the uppermost layer. Similarly, the bed should be stirred from time to time to give even drying and to prevent the bottom paddy from being overdried.

The best solar dryer designs for grain drying incorporate high air flow rates with only moderate increases in temperature. Hence, this is an obvious area for the use of forced convection dryers. Most of the research and development in this field has been carried out in North America, reflecting the economic importance of cereals in this region.

One type of forced convection solar dryer is shown in figure 12. This dryer type was originally designed to improve the drying rate of grain contained in conventional bins. The roof and sun-facing wall of the bin are converted into a solar collector by painting them black to enhance the absorption of solar radiation. An air duct is formed by fixing wooden panels under the roof and inside the sun-facing wall. The heat absorbed by the black surface is transmitted by conduction to the air within the duct, thus raising its temperature. A fan at the base of the air duct draws the warm air from the duct, and forces it through the drying bed. After passing through the grain, the exhaust drying air passes through the chimney in the roof.

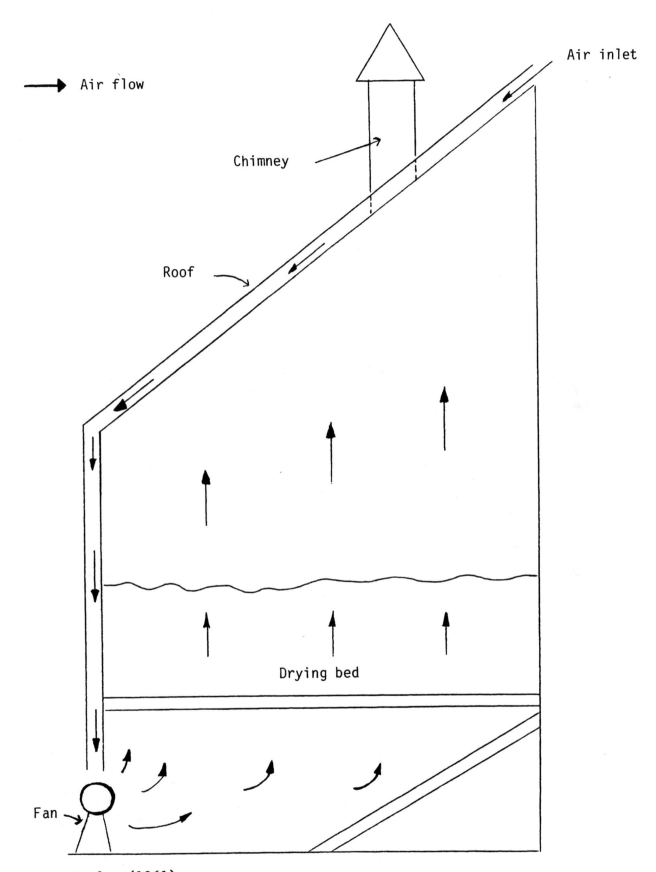

Air flow

Air inlet

Chimney

Roof

Drying bed

Fan

Source: Buelow (1961)

Figure 12

A forced convection solar dryer

Many other designs are available, details of which can be found in the references listed at the end of this document.

Where forced air convection drying appears to be a promising solution to a drying problem, the relative merits of solar energy systems and conventional artificial drying systems should be investigated. Artificial dryers require a supply of energy to heat the drying air and to drive the fan. Solar dryers require energy to drive the fan, and to justify this cost, the collector must be efficient enough to heat the air to the required temperature. Both types of systems must therefore be carefully costed. The possibility of the hybrid system of artificial drying with solar enhancement should also be considered.

To reduce the dependence of forced convection solar dryers on conventional fuels, photovoltaic systems are being developed which convert sunlight into electricity. This electricity could then be used to drive the fans. This technology is still in its infancy and has limited uses. When developed, and production costs of photovoltaic panels come down, it should provide a clean, cheap source of electricity which would have widespread uses, including powering fans in forced convection solar dryers.

III. Packaging and storage

Cereal grains are well suited to handling and storage in bulk, and can be packed in jute or cotton sacks or in bulk storage containers. During storage, deterioration can be caused by heat, moisture, moulds, insects, mites, rodents, and birds.

Deterioration by heat can be minimised by storing in the shade in a well ventilated area. If a storage building is used, excessive heat gain can be prevented by reflection of solar radiation, correct orientation, shading of walls, insulation, and controlled ventilation.

In storage buildings the roof, walls, doors, and ventilation openings should exclude wind-blow rain. If water absorption from the ground below is a problem, the floor must be supplied with a damp-proof course or the packaged goods should be stacked on pallets, boards, or poles giving 10 cm floor clearance, or moisture-proof sheets.

Rodents and birds can be excluded from buildings by sealing all unnecessary openings, screening the ventilation openings and minimising gaps around doors. Poisoned baits can be used to eliminate rodents but should be used very carefully, preferably by experienced personnel.

Insects and mites can be killed by effective fumigation with a suitable gas such as phosphine or methyl bromide. Stacked produce can be fumigated under gas-proof sheets if these are properly used. Fumigation of the building itself can also kill pests throughout the interior of the building, but it is not always easily and effectively achieved. Insects and mites can also be controlled by treatment with contact pesticides, especially where these can be admixed with dried products, such as grains and pulses, for which there are approved and acceptable dosage rates. Only those pesticides recommended for such purposes should be used. Sacks should be stacked tidily and methodically so that the stacks can be kept clear of walls without collapsing.

Each stack of grain should be bordered by a clear gangway, and an access to the top of the stack should be provided. This is essential for inspection and pest control procedures. If necessary, ladders should be provided for access to the top of the stacks.

In building the stack, the bags need to be kept off the floor by placing them over stacking pallets or regularly spaced poles. The latter may not be necessary if the floor is completely moisture-proof, although they may still be useful in case of minor flooding.

The bags should be laid regularly as they are stacked, and should be bonded together by alternating the stacking pattern in each layer. Tight stacking is usually recommended provided that the grain is dry. Well-built stacks can include 30 to 40 layers of bags in sufficiently high stores.

Stacks can be completely covered with gas-proof sheets for fumigation. The sheets must be in good condition and held down firmly close to the floor by chains or narrow sand bags. For protection against reinfestation after fumigation, the stacks can be sprayed layer-by-layer, as they are built, with a locally approved semi-persistent contact insecticide. Subsequent respraying of the exposed stack surfaces will prolong the protection considerably but not indefinitely. Spraying the exposed surface (capping sprays) without the initial layer-by-layer application is often a waste of time and money. An alternative to spraying treatments consists in leaving the gas-proof sheets in place after fumigation as a physical barrier against insects. With this method, use of insecticides can be restricted to a spray or powder treatment of the floor where the edges of the protective sheets are laid. Excellent, long-term protection can be obtained by this procedure so long as the grain remains dry and cool (i.e. without significant heating above the ambient mean temperature). Caution is necessary, however, in high altitude areas where the daily fluctuations of ambient temperatures may be relatively large. In all situations, good warehouse management and regular inspection for damage to the sheets are essential. Monitoring arrangements are also advisable for the early detection of any infestation problems that could arise due, for example, to a fumigation failure or reinfestation through a torn sheet. The use of transparent, light-weight polyethylene sheets as permanent fumigation covers is usually less costly and more effective than other methods, and their transparency also permits inspection through the sheets.

Bag stacks in a store with controllable ventilation can also be effectively protected against reinfestation after fumigation by regularly spaced insecticide treatments. However, these must be applied sufficiently frequently and at the best time. A daily treatment at dusk is recommended. This method is particularly cost-effective when use is made of insecticides such as Dichlorvos. On the other hand, many of the commonly recommended alternatives, especially the natural and synthetic pyrethroids are usually considered rather too costly for daily application in grain stores. A final possibility, for stores that can be made effectively gas-tight and are permanently screened to prevent the entry of insects, is total store fumigation. This method allows a long-term protection of the grain, but requires a type of store and storage management which is usually impractical where stocks are being moved in and out at frequent intervals.

These same principles apply also to rural grain storage. If modern pesticides are not available or cannot be safely used, there is a need to make the best possible use of local traditional methods such as airtight storage. In many rural locations dried grain is stored in pits dug in the ground. This practice cannot be recommended if there is any possibility of dampness encroaching through the soil. The pit should be lined with an impervious material where possible.

Maize can be both dried and stored in a crib. To reduce insect infestation, the crib and the ground beneath it should be thoroughly cleaned before loading. Furthermore, any grain residues that cannot be of immediate use should be burnt. Spray treatment after cleaning, using an approved semi-persistent contact insecticide, is a useful further measure. If the cobs are to be stored longer than 3 to 4 months, they should be treated, layer by layer, as the crib is loaded, with a locally approved grain-protectant spray or powder. Alternatively, and more effectively, the grain can be protected by shelling the cobs as soon as they are dry (13 per cent moisture content) and admixing the grain protectant more thoroughly. This permits the effective use of substantially reduced dosage rates for most insecticides.

Other grains which have been adequately dried can also be stored in a crib. The walls should be meshed or woven compactly enough to prevent the threshed grains from falling through.

IV. Extension work activities for grain drying

The extension of grain drying requires a different approach from the other commodities discussed. Fish, fruit and vegetables contain over 80 per cent water in the fresh stage and this fraction has to be reduced to 20 per cent or less during drying. The drying requirement for cereals is much lower since most grains are harvested with a moisture content below 20 per cent. A distinction should be made between those cereals which require little if any secondary drying after harvest and those such as rice which require drying to ensure safe storage.

The first group includes most of the cereals grown in temperate or dry tropical climates such as wheat, maize and millet. Any secondary drying of these crops has to be considered largely as an extension of good storage practice.

The second group is typified by rice which is often harvested in hot humid conditions which are not conducive to easy sun drying. This group requires a more rigorous approach and any drying problems should be tackled separately, though not in isolation, from storage.

IV.1 Location

Since drying is best considered in conjunction with storage, the most suitable location to develop and demonstrate any improved drying technique will be at a centre involved in cereal storage. It may be possible to work in conjunction with a storage extension project. Of all the commodities and processes, grain storage receives more attention than most because of the importance of this food group. There is therefore likely to be an institute or agricultural extension agency working in this field within reach which can be consulted. Any storage project would of course be able to supply extension workers to study your methods.

IV.2 Current practices

The cultivation and handling of grains is an established practice in most countries and it is likely that the methods used by the farmer to harvest and store the grain, together with any problems experienced, are well documented. This, of course, does not mean that solutions have been found to all of the problems.

The current drying requirements should be considered. If only an occasional batch is endangered, due to say freak showers, then the simplest solution might be to devise a cover which can be used with sun drying. Where maize is taking too long to dry in a crib, it might be possible to redesign the crib to allow a better air flow. This may entail the use of alternative construction materials or methods. A simple solution might be to recommend a different method of filling the crib so that the grain is less densely packed, and thus more moisture is allowed to escape. However, it should be remembered that looser packing may permit easier entrance for insect and pests.

Where there are more substantive drying demands, such as in the case of rice, the use of solar dryers to improve drying rates could be considered.

IV.3 Demonstration units

Ideally, these should be constructed side by side with examples of grain stores, in consultation with storage technicians. This will permit an integrated demonstration showing the advantages of both good drying and storage practices.

Decide which drying techniques are going to be demonstrated and build the necessary units. Most of the devices will be quite straightforward with the possible exception of the solar paddy dryer. Become familiar with the necessary construction techniques for the solar dryer and find out whether the necessary materials and skills are to be found in the rural sector. The design may have to be modified until a functional device, which can be built cheaply and with minimum effort, is obtained. Drying trials should be carried out until the operation of the dryer is mastered and its effectiveness is demonstrated.

In conjunction with people knowledgeable in storage matters, the quality of all dried cereals prepared by the different methods should be evaluated.

Some simple costing exercises should be done to determine what the additional costs of the improvements will be to the farmer and whether the benefits (improved quality, reduced losses, etc.) justify these costs.

IV.4 Involvement of extension workers

Having tried and tested modifications of existing drying techniques or developed new methods, this information can now be disseminated to extension workers. Ideally, a training programme in drying should be carried out in conjunction with a session on storage. The most likely trainees to benefit from this combined approach will be extension workers working on grain storage projects.

During this training programme, the extension workers should be made aware of:

- correct field drying practices;
- the vagueries or possible inadequacies of sun drying;
- the damage caused to incorrectly dried grain by moulds;
- the relative merits of threshing and drying the grain on the head;
- simple improvements to sun drying;
- the potential for solar drying;
- the construction materials and correct use of any drying device;
- the additional cost of any improvement;
- correct grain storage practice;
- the damage caused by insects, rats, etc. to stored grains.

At the end of the training programme, the extension worker should be in a position to advise farmers of the most appropriate techniques to use at all stages from harvesting to storage.

BIBLIOGRAPHY

I.   FOOD PROCESSING AND GENERAL DRYING

Baron, C. (ed.): Technology, employment and basic needs in food processing in developing countries (Oxford, Pergamon Press, 1980).

Budyko, M.I.: The heat balance of the earth's surface. English translation by N. Stepanova (Washington, DC., United States Department of Commerce and Weather Bureau, 1958).

Esmay, M.L.; Hall, C.W. (eds): Agricultural mechanisation in developing countries. (Tokyo, Sjin-Norinsha Co., 1974).

ILO: Appropriate technology for employment generation in the food processing and drink industry in developing countries, Technical Report III, Second Tripartite Technical Meeting on Food Processing and Drink Industries (Geneva, 1978).

Van Arsdel, W.B.; Copley, M.J.; Morgan, A.I.: Food dehydration, Vol. I and II. (Westport, Avi Publishing Co., 1973).

Van Loesecke, H.W.: Drying and dehydration of food. (New York, Reinhold Publishing Co., 1957).

Van Wylen, G.J.; Sonntag, R.E.: Fundamentals of classical thermodynamics, Second edition. (New York, John Wiley and Sons, 1976).

Yaciuk, G. (ed.): Food drying, Proceedings of a workshop held at Edmonton, Alberta, 6-9 July 1981 (Ottawa, International Development Research Centre, 1982).

II. SOLAR DRYING

Anon: How to make a solar cabinet dryer for agricultural produce, Do-it-Yourself Leaflet No. L6 (St. Anne de Bellevue, Canada, Brace Research Institute, 1965).

Bolin, H.R.; Salunkhe, D.K.: "Food dehydration by solar energy - Critical reviews", in Food Science and Nutrition, 1982, vol. 16 (4), pp. 327-354.

Buelow, F.H.: "Drying grain with solar heated air", in Quarterly Bulletin vol. 41 (2), pp. 421-429 (East Lansing, Michigan Agricultural Experimental Station, 1958).

-----: Drying crops with solar heated air, Proceedings of a UN Conference on New Sources of Energy (Rome, United Nations, 1961).

-----: "Corrugated heat collectors for crop drying", in Sun at Work, 1981, vol. 4, pp. 8-9.

Harigopal, V.; Tonapi, K.V.: "Technology for villages - Solar dryer" in
    Indian Food Packer, 1980, vol. 34 (2), pp. 48-49.

Lawand, T.A.: Solar dryers for farm produce, Technical report T4 (St. Anne
    de Bellevue, Canada, Brace Research Institute, 1963).

---: "A solar cabinet dryer", in Solar Energy, 1966, vol. 10 (4),
    pp. 158-164.

---: The operation of a large-scale solar agricultural dryer, Progress
    Report, Technical Report T33 (St. Anne de Bellevue, Canada, Brace Research
    Institute, 1967).

McDowell, J.: Solar drying of crops and foods in humid tropical climates,
    Report CFNI-T-7-73 (Kingston, Jamaica, Caribbean Food and Nutrition
    Institute, 1973).

Mendoza, E.P.: Performance of a low-cost solar crop dryer, Proceedings of
    an inter-regional symposium on solar energy for development, doc.
    B-7 (Tokyo, Japanese Solar Energy and Technology Association, 1979).

New Mexico Solar Energy Association: How to build a solar crop dryer (Santa Fe,
    New Mexico, NMSEA, 1978).

Szulmzyer, W.: "From sundrying to solar dehydration", in Food Technology, 1971,
    vol. 23, pp. 440-443.

Trim, D.S.: Solar crop dryers, Paper presented at the Arab Conference on
    Solar Energy Utilisation in Agriculture at Annan, Jordan, December 1982
    (London, Tropical Development and Research Institute, 1982).

Williams, B.: "Design of solar home food dryers", in Sunworld, 1980, vol. 4
    (6), pp. 195-196.

III. FISH

Burgess, G.H.; Cuttling, C.L., Lovern, J.A.; Waterman, J.J.: Fish handling
    and processing (New York, Chemical Publications Co., 1967).

Curran, C.A.; Trim, D.S.: Comparative study of three solar dryers for use
    with fish, Proceedings of an FAO Expert Consultation on Fish Technology
    in Africa held in Casablanca, Morocco (Rome, FAO, 1982).

Deng, J.C.; Chaw, K.V.; Baird, C.D. Heinis, J.J.; Perez, M. and Wu, L.: Drying
    seafood products with solar energy, Proceedings of the Second
    International Conference on Energy Use Management (Los Angeles, 1979).

Doe, P.E.; Ahmed, M. Muslenmuddin, M.; Sachithananthan, K.: "A polythene
    tent dryer for improved sun drying of fish", in Food Technology in
    Australia, 1977, vol. 29, pp. 437-441.

---: A polythene test fish dryer - A progress report, Proceedings of
    an international conference on Agricultural Engineering in National
    Development, doc. 79-12 (Selangor, Malaysia, University Pertanian, 1979).

FAO: The prevention of losses in cured fish, FAO Fisheries Report Paper
    No. 219 (Rome, 1981).

ILO: Small-scale processing of fish, Technical Memorandum No. 3 (Geneva, 1982).

Richards, A.H.: A polythene tent fish dryer for use in Papua New Guinea's Sepik River salt fish industry, Proceedings of a seminar on sun drying methodology (Colombo, Sri Lanka, National Science Council, 1976).

Rogers, J.F.; Cole, R.C.; Smith, J.D.; Barron, J.O.: An illustrated guide to fish preparation, Report G83 (London, Tropical Development and Research Institute, 1975).

Sachithananthan, K.; Trim, D.; Speirs, C.I.: A solar dome dryer for drying of fish, FAO Fisheries Paper (Rome, FAO, 1983).

Watanabe, K.: "An experimental fish drying and smoking plant on Volta Lake: Design, construction and economic considerations", in Tropical Science, 1975, vol. 17, No. 2, pp. 75-93.

Waterman, J.J.: The production of dried fish, Fisheries Paper No. 160, (Rome, FAO, 1976).

IV. VEGETABLES

Anon: Drying fruits and vegetable at home, Bulletin No. 555 (University of Idaho, 1975).

---: "Chilli drying, vegetable seeds drying", in Annual Report (Bhopal, India, Central Institute of Agricultural Engineering, 1980), pp. 51-52.

Best, R.: Cassava drying. (Cali, Colombia, Centro Internacional de Agricultura Tropical, 1979).

Jackson, T.H.; Mohammed, B.B.: Sun drying of fruits and vegetables, Agricultural Services Bulletin No. 5 (Rome, FAO, 1969).

Mantosudirvo, S.; Kurisman, I.; Taragan, I.: Improvement of solar drying technique in post-harvest technology - A study of onion drying in Indonesia, Proceedings of an inter-regional symposium on solar energy for development, Paper B-10 (Tokyo, Japanese Solar Energy and Technology Association, 1979).

Shaw, R.: "Solar drying potatoes", in Appropriate Technology, 1981, vol. 7, No. 4, pp. 26-27.

Start Clark, C.: "Solar food drying: A rural industry", in Renewable Energy Review Journal, 1981, vol. 3, No. 1, pp. 23-26.

Trim, D.S.; Ko, H.Y.: "Development of a forced convection solar dryer for red peppers", in Tropical Agriculture, 1982, vol. 59, No. 4, pp. 319-323.

V. FRUITS

Anon: Drying fruit and vegetables at home, Bulletin No. 555 (University of Idaho, 1975).

Bhatia, A.K.; Gupta, S.L.: "Solar dryer for drying apricots", in Research and Industry, 1976, vol. 21, No. 9, pp. 188-191.

Bolin, H.R.; Petrucci, V.; Fuller, G.: "Characteristics of mechanically harvested raisins produced by dehydration and by field drying", in Journal of Food Science, 1975, vol. 40, p. 1036.

Bolin, H.R.; Stafford, A.E.: "Effect of processing and storage on provitamin A and vitamin C in apricots", in Journal of Food Science, 1974, vol. 39, p. 1035.

Bolin, H.R.; Stafford, A.E.; Huxsoll, C.C.: "Solar heated fruit dehydrator", in Solar Energy, 1978, vol. 20, pp. 289-291.

Bowrey, R.G.; Buckle, K.A.; Haney, I.; Pavenayotin, P.: "Use of solar energy for banana drying", in Food Technology in Australia, 1975, vol. 32, No. 6, pp. 290-291.

Cheema, L.S.; Riberio, C.M.C.: Solar dryers of cashew, banana and pineapple, Proceedings of a conference on "The sun: Mankind's future source of energy" (Parkville, Australia, International Solar Energy Society, 1978), pp. 2075-2079.

FAO: Date production and protection, Plant Production and Protection Paper No. 35 (Rome, FAO, 1983).

Jackson, T.H.; Mohammed, B.B.: Sun drying of fruits and vegetables, Agricultural Services Bulletin No. 5 (Rome, FAO, 1969).

Pablo, I.S.: The practicality of solar drying of tropical fruits and marine products for income generation in rural areas, Proceedings of a UNESCO Solar Drying Workshop (Manila, Philippines, Bureau of Energy Development, 1978).

Umaro, G.G.; Ikramov, A.I.: "Features of the drying of fruits and grapes in solar radiation drying apparatus", in Geliotekhnika, 1978, vol. 14, No. 6, pp. 55-57.

VI. GRAINS

Boshof, W.: Drying and conserving maize in the humid tropics, Tropical Stored Products Information No. 36 (London, Tropical Development and Research Institute, 1978).

Brooker, D.B.; Bakker-Arkema, F.W.; Hall, C.W.: Drying of cereal grains. (Westport, Conn., Avi Publishing Co., 1974).

Exell, R.H.B.; Kornsakoo, S: "A low cost solar rice dryer", in Appropriate Technology, 1978, vol. 5, No. 1, pp. 23-24.

---;---: "Solar rice dryer", in Sunworld, 1979, vol. 3, No. 3, p. 75.

Exell, R.H.B.: "Basic design theory for a simple solar rice dryer", in Renewable Energy Review Journal, 1980, vol. 1, No. 2, pp. 1-14.

Foster, G.H.; Peart, R.M.: Solar grain drying: Progress and potential, Agricultural Information Bulletin No. 401 (Washington DC., Department of Agriculture, 1976).

Hall, D.W.: Handling and storage of food grains in tropical and sub-tropical areas, Agricultural Development Paper No. 90 (Rome, FAO, 1970).

ILO: Small-scale maize milling, Technical Memorandum No. 7 (Geneva, 1984).

Kent, N.L.: Technology of cereals. (Oxford, Pergamon Press, 1966).

O'Kelly; E.: Processing and storage of foodgrains by rural families, Agricultural Services Bulletin (Rome, FAO, 1979).

Ozisik, M.N.; Huang, B.K.; Toksoy, M.: "Solar grain drying", in Solar Energy, 1980, vol. 24, pp. 397-401.

Wumbleski, E.M.; Catania, P.J.: Solar grain drying in Saskatchewan, Canada, Proceedings of a conference "The sun: Mankind's future source of energy" (Parkville, Australia, International Solar Energy Society, 1978).

Printed in the United States
48169LVS00001B